FOUNDATIONS OF QUANTUM MECHANICS SINCE THE BELL INEQUALITES

SELECTED REPRINTS

EDITED BY
L.E. BALLENTINE

PUBLISHED BY AMERICAN ASSOCIATION OF PHYSICS TEACHERS

1988

Foundations of Quantum Mechanics Since the Bell Inequalities

Published by:

American Association of Physics Teachers
5112 Berwyn Road
College Park, MD 20740, U.S.A.

Cover: The diagram on the cover depicts an experiment for testing Bell's inequalities. It appears on page 95 in "Experimental Test of Bell's Inequalities Using Time-Varying Analyzers" by A. Aspect, J. Dalibard and G. Roger. Cover design by Rebecca Heller Rados.

ISBN #: 0-917853-33-4

Contents

Resource letter IQM-2: Foundations of quantum mechanics since the Bell inequalities

L. E. Ballentine
Department of Physics, Simon Fraser University, Burnaby, British Columbia, Canada V5A 1S6

(Received 16 January 1987; accepted for publication 27 January 1987)

This Resource Letter provides a guide to the literature on the foundations of quantum mechanics over approximately the past 20 years. Topics covered include Bell's theorem, interpretation of the quantum state concept, the theory of measurement, and experimental tests of fundamental aspects of the quantum theory of matter and the electromagnetic field. The letter E after an item indicates elementary level of material of general interest to persons becoming informed in the field. The letter I, for intermediate level, indicates material of somewhat more specialized nature; and the letter A indicates rather specialized or advanced material. An asterisk (*) indicates those articles to be included in an accompanying Reprint Book.

I. INTRODUCTION

The purpose of this Resource Letter is to review the developments in the foundations of quantum mechanics that have taken place during approximately the past 20 years, a period in which Bell's theorem has been the most spectacular development although by no means the only one. Another convenient reference point is the Resource Letter IQM-1 by DeWitt and Graham (Ref. 14) published in 1971. For the most part I shall not include papers that are listed in that earlier Resource Letter.

A glance at IQM-1 reveals that the Copenhagen interpretation was the dominant interpretation during that earlier period. It is doubtful whether that is true any longer. But in order to make such a statement it is necessary to specify what is meant by the "Copenhagen interpretation." I shall use the term in its popular, but not necessarily historically accurate sense, as including the following propositions:

(a) The state vector provides a complete description of an individual system (that is, the views of Bohr rather than of Einstein in their famous controversy);

(b) The state vector evolves according to the Schrödinger equation while the system is isolated, but changes discontinuously during measurement to an eigenstate of the observable that is measured (Von Neumann's "*projection postulate*"). Of course this does not fully characterize the Copenhagen interpretation. In particular, Born's statistical interpretation of the wavefunction is also an essential ingredient, but it is either a postulate or a consequence of almost every interpretation, so it does not distinguish one from another. Some will not like my terminology. L. Rosenfeld has said (quoted in Ref. 21), "We in Copenhagen do not like (the phrase Copenhagen interpretation) at all. Indeed, this expression was invented, and is used by people wishing to suggest that there may be other interpretations ···, namely their own muddled ones." Thus not only is the interpretation of QM controversial, but the existence of a controversy (which requires at least two interpretations) is controversial! There is no point in arguing about labels, and I shall employ the label "Copenhagen interpretation" to refer to the well-known ideas listed above.

One of the most interesting features of the newer interpretations is, in spite of their apparent diversity, a convergence on certain tenets that contradict the old Copenhagen interpretation. Ballentine (Ref. 18, *Statistical ensemble interpretation*) draws his philosophical inspiration from Einstein, criticizes the "Copenhagen interpretation," and rejects (a) and the projection postulate of (b). Stapp (Ref. 21) draws his philosophical inspriation from Bohr and calls his paper "The Copenhagen Interpretation," but he rejects several of Heisenberg's ideas, as well as the interpretation of which (b) is the corner stone (which, however, he calls "Absolute Ψ"). Moreover, it appears that Ballentine and Stapp are in substantial agreement about the concept of "state" (see Ref. 36). In order to distinguish it from the old Copenhagen interpretation above, I refer to Stapp's views as the "*neo-Copenhagen interpretation*."

None of the principal modern interpretations (Sec. V A) accept the consequence of Von Neumann's projection postulate, the so-called "reduction" or "collapse" of the state vector, as a real physical process. This rejection is necessary because the result of a consistent inclusion of the measurement apparatus in the system has been proved to be strictly incompatible with the result of the projection postulate (Sec. V D). The projection postulate is at best an approximation, valid under limited circumstances, and so has no place as an axiom.

Bell's theorem is regarded by many people as one of the most profound and startling developments in modern physics, affecting not only the interpretation of QM, but ultimately our whole world view. But the awareness of its significance was slow to develop, as may be seen from the graph (Fig. 1) of citations of Bell's original paper.

Bell's work is a continuation, after nearly 30 years, of the

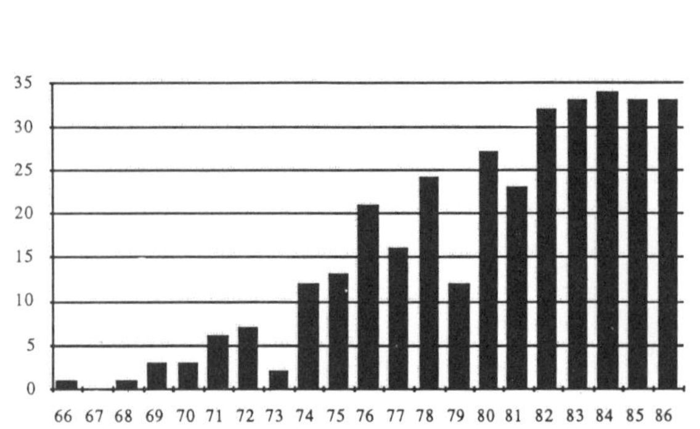

Fig. 1. Annual citations in *Science Citation Index* of Bell's paper (Ref. 53).

famous paper by Einstein, Podolsky, and Rosen. The essence of their argument was to consider a pair of particles which have been prepared in an eigenstate of total momentum $p_1 + p_2$ and relative position $q_1 - q_2$. If one measures p_1, one can deduce the value of p_2; if one measures q_1, one can deduce the value of q_2. Invoking a postulate of *locality*, of a type suggested by special relativity, that there is no action-at-a-distance between the particles, EPR inferred that the values of p_2 and q_2 could not have been created by the measurement, but must have been preexistent *elements of reality*. Since no quantum state can describe simultaneous values for noncommuting variables like position and momentum, EPR concluded that the description of reality by QM states is incomplete.

Most of the published responses to EPR are only of historical interest. I mention only that of Margenau (Ref. 51), who pointed out that the EPR paradox highlights the need to eliminate Von Neumann's projection postulate. That conclusion, which has subsequently been reinforced by more powerful arguments, was relevant to the original form of the EPR argument. But the EPR argument can be reformulated without the projection postulate, as is done above, and it still retains its content. Any paradoxical aspects of the conclusions of the EPR and Bell arguments do not depend upon the notion of state-vector-reduction.

Bell's inequalities, in their most general form, are restrictions on the degree of correlation between two separated detectors, and are derivable from a locality postulate similar to that of EPR. *Bell's theorem* is that some predictions of QM violate these inequalities, even though there may be *no apparent mechanism* for action-at-a-distance. There is some confusion about the relation of Bell's theorem to the EPR argument, it being sometimes said that Bell's conclusion contradicts that of EPR. In fact Bell's argument is fully compatible with that of EPR, of which it is an extension. This can most easily be seen in the simplified presentation of Mermin (Ref. 70). One considers a source of pairs of spin $\frac{1}{2}$ particles in the singlet ($J = 0$) state. For each particle one chooses to measure the spin component in one of three directions, which form angles of 120°. A contradiction of Bell's inequality can be obtained from the combination of two predictions of QM: (i) For those pairs whose spins are measured along the same direction the results will be perfectly (oppositely) correlated, (ii) If one averages over all nine possible choices for the pair of directions then the average correlation for all pairs of particles is zero. EPR in effect used only the information (i), and hence Bell's result are more powerful. Although Bell's theorem casts serious doubt about the validity of the locality postulate used by EPR, it does not falsify their conclusion that QM is *incomplete* (in their technical sense of the word). Moreover, that conclusion seems to be confirmed by an independent argument (Ref. 115).

Many papers have been written about the significance and interpretation of Bell's theorem. What is the precise nature of the *locality* postulate that must be used in the derivation? Are more assumptions needed than merely *locality*? If indeed we must give up *locality*, what kind of nonlocality must we have, and how will this affect our world view? It is easy enough to invent an interpretation that admits nonlocality (allowing causality to propagate in both the forward and backward light cones is perhaps the easiest), but so far such attempts have not been very illuminating. Perhaps what is needed is not an explanation of nonlocality, but an explanation of locality. Why, if locality is not true, does it work so well in so many different contexts?

The greatest change since the publication of Resource Letter IQM-1 is the increase in the number of *experiments* which test or illustrate the fundamental aspects of QM. IQM-1 lists only two experiments, one of which is only a proposal. Now the list includes: tests of the Bell inequalities; measurement of the electron momentum distribution in atomic hydrogen; interference experiments in which neutron beams are split, separated by centimeters, spin-flipped, and recombined coherently (confirming the superposition principle and contradicting "collapse" of the state vector); electron holography to make wavefunction phase fronts visible; observation of the Kapitza–Dirac effect (diffraction of matter by light, instead of the familiar diffraction of light by matter); and photon correlation measurements which are able to distinguish between quantum electrodynamics and semiclassical radiation theories. No longer can it be claimed that the interpretation of QM is a dull subject for endless debate by philosophers and armchair physicists. The foundations of QM has become what it always should have been, an exciting subject at the heart of mainstream physics.

II. CONFERENCE PROCEEDINGS

1. **Proceedings of the International School of Physics "Enrico Fermi,"** Course 49 (Varenna, 29 June–11 July 1970), "Foundations of Quantum Mechanics," edited by B. d'Espagnat (Academic, New York, 1971). (A)
2. **Les Implications Conceptuelles de la Physique Quantique,** J. Phys. (Paris) Coll. **C-2** (March 1981). In French, except for the article b· Bell. (A)
3. **Foundations of Quantum Mechanics in the Light of New Technology.** proceedings of an international symposium held at the Central Research Laboratory of Hitachi, Ltd., Tokyo, 29–31 Aug. 1983. Edited by S. Kamefuchi (Physical Society of Japan, Tokyo, 1984). (A)
4. **Open Questions in Quantum Physics,** proceedings of a workshop held in Bari, Italy, May 1983. Edited by G. Tarozzi and A. van der Merwe (Reidel, Dortrecht, 1985). (A)
5. **Fundamental Questions in Quantum Mechanics,** proceedings of a conference at Albany, NY, 12–14 April 1984. Edited by L. M. Roth and A. Inomata (Gordon and Breach, London, 1986). (I,A)
6. **Symposium on the Foundations of Modern Physics: 50 years of the Einstein–Podolsky–Rosen Gedankenexperiment,** proceedings of a conference at Joensuu, Finland, 16–20 June 1985, edited by P. Lahti and P. Mittelstaedt (World Scientific Publishing Co., Singapore, 1985). (I,A)
7. **New Techniques and Ideas in Quantum Measurement Theory,** edited by D. M. Greenberger, Annals NY Acad. Sci. **480** (1986). Proceedings of a conference at New York, 21–24 Jan. 1986. (I,A)

III. TEXTBOOKS AND EXPOSITIONS

Few textbooks deal adequately with the problems of interpretation of QM. Many give a perfunctory rehash of ideas circa 1930; others remain silent. A notable exception is the following book.

8. **Modern Quantum Mechanics,** J. J. Sakurai (Benjamin/Cummings, Menlo Park, California, 1985). This is the only textbook that discusses Bell's inequalities. (I)

There are some specialized expositions that deal with these problems.

9. **A survey of Hidden-Variable Theories,** F. J. Belinfante (Pergamon, Oxford, 1973). The scope of this book is wider than the title would suggest. A "hidden-variable model" is broadly interpreted as any dynamical or statistical model that is intended to illustrate some or all of the features of QM. Belinfante discusses the content of various theorems that pertain to such models, often in terms much clearer than those of the original authors. The general theorems are illustrated by detailed studies of specific models. Part III of the book is devoted to the derivation and explanation of Bell's inequalities, and to detailed examination of their relation to specific models and to experiments. (I)
10. **The Philosophy of Quantum Mechanics: The Interpretations of Quantum Mechanics in Historical Perspective,** M. Jammer (Wiley, New York, 1974). A good account of the historical development of various interpretations of QM (mainly the older ones, but Bell's theorem is included). However, the reader should be aware that this is primarily a *history* of ideas, which are faithfully represented even when fallacious, and little criticism is provided beyond that which already exists in the literature. (I)
11. **Measurements and Time Reversal in Objective Quantum Theory,** F. J. Belinfante (Pergamon, Oxford, 1975). The purpose of this book is "primarily to give a reply to scientific papers written by a large number of authors," consequently it is not a self-contained introduction to the subjects of its title. But the author's personal views of those subjects are interesting and often helpful. (A)
12. **Conceptual Foundations of Quantum Mechanics,** B. d'Espagnat (Benjamin, Reading, MA, 1976). The topics include: a summary of relevant aspects of the formalism of QM; nonseparability (the EPR–Bell problem); measurement theories; and a discussion of philosophical views. (A)
13. **Chemistry, Quantum Mechanics and Reductionism,** H. Primas (Springer, Berlin, 1981). This book contains a remarkable combination of correct and contestable statements. It is *not* recommended for teaching, but readers who can excerise independent critical judgement may find it both interesting and challenging. In addition to a wealth of references, it contains good descriptive summaries and critiques of many interpretations. Ensemble interpretations receive little attention because of the author's prior belief that only an individual interpretation can supply his desiderata. He apparently believes that the "insoluble problem of measurement" (discussed under items 43 and 49), which bedevils all individual interpretations, can be resolved by means of superselection rules in the W^*-algebra formalism. That belief (or hope) is not well founded. (A)

IV. PREVIOUS REVIEWS

14. **Resource Letter IQM-1 on the Interpretation of Quantum Mechanics,** B. S. Dewitt and R. N. Graham, Am. J. Phys. **39**, 724–738 (1971). (E,I,A)
15. **"Quantum Mechanics Reality and Separability,"** F. Selleri and G. Tarozzi, Riv. Nuovo Cimento **4**, (2) 1–53 (1981). (I)
16. **Quantum Theory and Measurement,** edited by J. A. Wheeler and W. H. Zurek (Princeton U.P., Princeton, NJ, 1983). A reprint collection with a large supplementary bibliography. (E,I,A)
17. **"Bell's theorem: Experimental Test and Implications,"** J. F. Clauser and A. Shimony, Rep. Prog. Phys. **41**, 1881–1927 (1978). This is an excellent review of all subjects related to Bell's inequalities up to that date. (I,A)

V. CURRENT RESEARCH TOPICS

A. The principal interpretations of quantum mechanics

1. Statistical ensemble interpretation

* 18. **"The Statistical Interpretation of Quantum Mechanics,"** L. E. Ballentine, Rev. Mod. Phys. **42**, 358–381 (1970). It is argued that the quantum state function should be regarded as a description of a conceptual ensemble of similarly prepared systems, rather than as the complete description of an individual system. The opposite assumption is untenable because it leads to unsolvable difficulties in the theory of measurement. (I,A)

19. **"Einstein's Interpretation of Quantum Mechanics,"** L. E. Ballentine, Am. J. Phys. **40**, 1763–1771 (1972). Einstein's view of QM and the nature of his critical arguments are reviewed. Certain widespread misconceptions about his views are corrected. (I)

2. Neo-Copenhagen interpretation

20. **"S-Matrix Interpretation of Quantum Theory,"** H. P. Stapp, Phys. Rev. D **3**, 1303–20 (1971). Described by its author as "an elaboration upon the Copenhagen interpretation," this paper includes among its basic precepts: that the physical world must be separated into two parts, called the observed and the observing system (Stapp has subsequently come to view this as a serious limitation of the CI); and that the probabilities in the theory are probabilities of response of the measuring devices. "Collapse of the wave function" does not occur as a physical process, but only corresponds to a "change in the set of specifications on the preparation of the observed system." This paper is the first of many in which Stapp attempts to develop an interpretation of QM that is satisfactory in light of Bell's theorem. The appeal of this approach is limited by the unnaturalness of conceiving all processes as being essentially scattering processes. (A)

* 21. **"The Copenhagen Interpretation,"** H. P. Stapp, Am. J. Phys. **40**, 1098–1116 (1972). In attempting to save "the Copenhagen interpretation" the author radically revises what is often, rightly or wrongly, understood by that term. That interpretation in which Von Neumann's "reduction" of the state vector in measurement forms the core is rejected, as are Heisenberg's subjectivistic statements. The very "pragmatic" (one could also say "instrumentalist") aspect of the interpretation is strongly emphasized. Readers will be divided as to whether this is a virtue or a fault. (I)

3. Consistent histories interpretation

22. **"Consistent Histories and the Interpretation of Quantum Mechanics,"** R. B. Griffiths, J. Stat. Phys. **36**, 219–272 (1984). The usual formula for probabilities of single events is generalized to yield conditional probabilities for selected sequences of events at several different times. A sequence of events must satisfy certain conditions, which are well defined within the formalism but are rather difficult to grasp intuitively, in order to qualify as a "consistent history" to which a probability may be assigned. The state function plays a much less prominent role than in other interpretations and there is no physical process of "collapse." There is no need to distinguish conceptually between the observed and the observing parts of the system. (A)

23. **"Quantum Interpretation Using Consistent Histories,"** R. B. Griffiths, in Ref. 5, pp. 211–223 (1986). A less technical account of the concepts and results of the previous paper. (I)

4. Many worlds interpretation

This bizarre interpretation was motivated by the problem of quantum cosmology. The Copenhagen interpretation requries that there be an "observer" outside of the system who performs "measurements" on the system, and so it cannot be applied to the universe as a whole. The Statistical ensemble interpretation has no such limitation, and if applied to the whole universe could, in principle, yield a result such as, "The probability of a cosmos like ours developing out of the initial Big Bang is 42%." Such a prediction is not very useful, since it could be tested only by repeating the Big Bang a large number of times. The MWI postulates that the universe is described by a unique state vector. There is no "reduction" of the state vector, in measurement or at any other time, but whenever the state vector takes the form of a sum of orthogonal state vectors then each term is assumed to correspond to a separate world, or branch, of the universe. Appearances are saved because it is shown that an observer in one branch cannot detect any of the other branches. It was even claimed that "The mathematical formalism of quantum theory is capable of yielding its own interpretation." (That claim is refuted in Refs. 25 and 26.) Although the evolution of the universal state vector is deterministic, there is in principle no way to account for the particular combination of branches that constitutes ones personal history. Thus the predictive power of the MWI is the same as that of the Statistical ensemble interpretation.

24. **"The Many-Worlds Interpretation of Quantum Mechanics,"** edited by B. S. DeWitt and N. Graham (Princeton U.P., Princeton, NJ, 1973). Reprints of the original papers by H. Everett, J. A. Wheeler, B. S. DeWitt, L. N. Cooper, D. vanVechten, and N. Graham. (A)

25. **"Can the Statistical Postulate of Quantum Theory Be Derived?—A Critique of the Many-Universes Interpretation,"** L. E. Ballentine, Found. Phys. **3**, 229–240 (1973). (The MWI was originally called the "Many-Universes" interpretation, before being more appropriately renamed by its proponents.) The author points out that the definition of the "branches" of the universe in the MWI is dependent upon the choice of representation, and is therefore ambiguous. He also argues that the bizarre assumptions of the MWI are neither necessary nor sufficient to derive the Born statistical formula of QM. (A)

26. **"Quantum Theory and Cosmology,"** C. J. S. Clarke, Philos. Sci. **41**, 317–332 (1974). This author also shows that "some additional structures" must be added to the theory in order to determine which states will determine the "branching." (A)

27. **"How Many Worlds?,"** R. A. Healey, Nous **18**, 591–616 (1984). A critical discussion of several versions of the MWI (A)

5. The quantum potential

When the Schrödinger equation is written in terms of the real amplitude and phase of the complex wavefunction Ψ, the phase equation has the form of the Hamilton–Jacobi equation with an additional "quantum potential." This H–J equation generates an ensemble of particle trajectories that reproduce the QM probability distribution $|\Psi|^2$. The quantum potential does not necessarily become small in regions where the wavefunction is very small, and it can give rise to an effective interaction between particles that do not tend to zero at large separations. Because of this apparent nonlocality, the quantum potential model was often regarded as a mathematical curiosity that lacked any physical significance. But it has acquired a new lease on life since Bell's theorem has shown that some kind of nonlocality must be a feature of all interpretations of QM.

28. **"Quantum Interference and the Quantum Potential,"** C. Philippidis, C. Dewdney, and B. J. Hiley, Nuovo Cimento **52**, 15–28 (1979). The quantum potential and the ensemble of particle trajectories are computed and illustrated for the two-slit interference pattern. (I)

29. **"On the Impossible Pilot Wave,"** J. S. Bell, Found. Phys. **12**, 989–999 (1982). In this perceptive and provocative paper Bell asks why so many prominent people took seriously Von Neumann's "proof" that QM is incompatible with a more detailed deterministic description, and why similar "proofs" have been advanced by other authors, when an explicit counterexample to all such "proofs" was published by Bohm in 1952, and the essential idea for Bohm's model was introduced by deBroglie in 1927. Among the morals of the story is, "Always test your general reasoning against simple models." (I)

30. **"Unbroken Quantum Realism, from Microscopic to Macroscopic Levels,"** D. Bohm and B. J. Hiley, Phys. Rev. Lett. **55**, 2511–4 (1985). The quantum potential interpretation is put forward as a counterexample to recent claims that QM implies any denial of physical realism. (I)

31. **"An Ontological Basis for the Quantum Theory,"** D. Bohm, B. J. Hiley, and P. N. Kaloyerou, Phys. Rep. **144**, 321–375 (1987). A quite complete account of the quantum potential theory, applied to both particles and fields. (I)

B. Nature of quantum state functions

32. **"On Explaining the Meaning of the Wave Function,"** G. D. Hoyt, Am. J. Phys. **36**, 368–9 (1968). Stresses that $|\Psi|^2$ is a probability density for the *system*, and is not a matter wave for the electron. In the example of the hydrogen atom it gives as much information about the proton as about the electron. (E)

33. **"Quantum Fact and Fiction: Clarifying Lande's Pseudo-paradox,"** J.-M. Levy-Leblond, Am. J. Phys. **44**, 1130–2 (1976). A. Lande had objected that deBroglie's relation $p = h/\lambda$ violates Galilean invariance, since the wavelength λ of a wave is the same in two frames of reference with relative velocity v, but the momentum p of a particle transforms as $p' = p + mv$. Levy-Leblond points out that, unlike a classical wave, the wavefunction in QM undergoes a phase transformation in addition to the transformation of coordinates ($x' = x + vt$, $t' = t$), and that this causes the deBroglie wavelength λ to transform just so as to maintain the relation $p = h/\lambda$. He stresses that the notion of "wave-particle duality" can be misleading because the wavefunction of QM is not a real wave. (I)
34. **"Probability Interpretation of Quantum Mechanics,"** R. G. Newton, Am. J. Phys. **48**, 1029–34 (1980). The author rejects the interpretation of the wavefunction as a real object in space, or as an expression of subjective knowledge, in favor of an objective statistical interpretation. He emphasizes that an ensemble interpretation is required for both pure and mixed states. (I)

*35. **"Solution to Two Paradoxes in the Quantum Theory of Unstable Systems,"** M. Bunge and A. J. Kalnay, Nuovo Cimento **77B**, 1–9 (1983). According to some people, all decay processes should be exponential. According to others, an unstable system under constant observation will never decay. Both of these incorrect conclusions are shown to be consequences of an incorrect interpretation of quantum states. (I)

36. **"What is a State Vector?,"** A Peres, Am. J. Phys. **52**, 644–650 (1984). A state vector is not a property of a single system, but rather of a *preparation procedure*. He quotes Ballentine (Ref. 18, "Statistical ensemble interpretation") and Stapp (Ref. 21, "Copenhagen interpretation") to show that both agree on this point. (I)

C. State preparation and state determination

*37. **"An Operational Interpretation of Nonrelativistic Quantum Mechanics,"** W. E. Lamb, Jr., Phys. Today **22**, (4) 23–28 (1969). Methods are described by means of which an arbitrary quantum state could, in principle, be prepared. Preparation is clearly distinguished from measurement. (I)

38. **"The Empirical Determination of Quantum States,"** W. Band and J. L. Park, Found. Phys. **1**, 133–144 (1970). (I)
39. **"A General Method of Empirical State Determination in Quantum Physics: Part II,"** W. Band and J. L. Park, Found. Phys. **1**, 339–343 (1971). Given a definite preparation procedure, how can one determine what state is actually being prepared? These two papers show how it can be done by measuring the averages, on a ensemble of similarly prepared systems, of a "quorum" of noncommuting observables. (I)
40. **"The Physics and the Semantics of Quantum Measurement,"** H. Margenau and J. L. Park, Found. Phys. **3**, 19–28 (1973). In the course of criticizing another paper, the authors provide a clear distinction between *state preparation* and *measurement*. (I)
41. **"Measurement of the Wigner Function,"** A. Royer, Phys. Rev. Lett. **55**, 2745–8 (1985). The Wigner function provides a phase space representation of a quantum state. Since it is not a probability, its operational significance has been unclear. Royer shows how, in principle, the Wigner function (and hence the state) can be experimentally determined for an ensemble of particles. (I)

D. Theory of measurement

Only papers that use the theory of measurement for the purpose of clarifying the interpretation of QM are included in this selection. Some papers on other aspects of the subject, such as the ultimate limits of measurements, are contained in Ref. 16. Reference to older works may be found in Ref. 14.

42. **"Alternative to the Orthodox Interpretation of Quantum Theory,"** P. Pearle, Am. J. Phys. **35**, 742–753 (1967). Descriptions of the measurement process are given, according to two interpretations (the "old Copenhagen" and "Statistical ensemble" interpretations, in the terminology of this Resource Letter). Possible (but not yet practical) experiments to distinguish between the two are described. (I)
43. **"Insolubility of the Quantum Measurement Problem,"** A. I. Fine, Phys. Rev. D **2**, 2783–7 (1970). The "problem" is whether a measurement interaction can leave the system (object + apparatus) in a mixed state, each component of the mixture displaying a definite value for the apparatus's indicator observable. It is shown that this is not possible. (Thus the "problem" is really not insoluble, as stated in the title, but rather it is answered negatively.) (A)
44. **"Is there a Quantum Measurement Problem?,"** P. A. Moldauer, Phys. Rev. D **5**, 1028–32 (1972). It is argued that the state vector, *without* Von Neumann's "reduction" hypothesis, correctly describes the results of measurements, including correlations between successive measurements. Hence there is no "measurement problem." (A)
45. **"There is a Measurement Problem: A Comment,"** A. I. Fine, Phys. Rev. D **5**, 1033 (1972). Fine agrees with the content of Moldauer's paper (Ref. 44), but points out that it involves a change from what had been the usual interpretation of QM. (A)
46. **"Approximate Measurement in Quantum Mechanics,"** A. Shimony, Phys. Rev. D **9**, 2321–3 (1974). The theorem of Ref. 43 is extended to include approximate measurements, and a technical defect in that proof is remedied. (A)
47. **"Quantum Measurements are Reversible,"** A. Peres, Am. J. Phys. **42**, 886–891 (1974). A simple example is given to show that the state of the total system, object plus measuring apparatus, does not change from a pure to a mixed state during the measurement process. But if one ignores the correlation of the apparatus to the object and attempts to describe the object alone, then it will appear to have a mixed state. (E)
48. **"Quantum Theory of Measurement Without Wave Packet Collapse,"** M. Cini, Nuovo Cimento **73B**, 27–56 (1983). The main points of this paper are the "projection postulate" should not be part of the theory and "wave packet collapse" does not exist as a separate physical process. However, the loss of coherence predicted by the "collapse" postulate is often an approximate consequence of the macroscopic nature of the measuring apparatus. This need not involve irreversibility. (A)

*49. **"What is the Point of the Quantum Theory of Measurement?,"** L. E. Ballentine, in Ref. 5, pp. 65–75 (1986). Proves, under very general conditions, that the interaction of the measured object and the measuring appartus must yield a final state of the system (object + apparatus) that is a *coherent superpostion* of *marcroscopically distinct* "pointer position" eigenvectors. Thus any interpretation of QM which requires a "reduction" of the state to an incoherent mixture of "pointer position" states (or to a single such state) is untenable. In particular, the old idea that "disturbances" during the measurement process cause such a "reduction" is proved false. (I)

E. Bell's theorem and its consequences

1. *Before Bell's theorem*

50. **"Can Quantum-Mechanical Description of Physical Reality be considered Complete?,"** A. Einstein, B. Podolsky, and N. Rosen, Phys. Rev. **47**, 777–80 (1935). Reprinted in Ref. 16. (E,A)
51. **"Quantum-Mechanical Description,"** H. Margenau, Phys. Rev. **49**, 240–2 (1936). Points out that many of the paradoxical features in the situation described by EPR cease to be paradoxical when one drops the assumption that a measurement projects that state of the measured system onto an eigenstate of the observable that is measured. (I,A)
52. **"Discussion of Experimental Proof for the Paradox of Einstein, Podolsky and Rosen,"** D. Bohm and Y. Aharonov, Phys. Rev. **108**, 1070–6 (1957). Reformulation of the argument in terms of the correlated spins, thus making experimental tests more feasible. (I,A)

2. *Bell's inequality*

*53. **"On the Einstein Podolsky Rosen Paradox,"** J. S. Bell, Physics **1**, 195–200 (1964). The derivation of the inequality in this seminal paper is appropriate only for the case of *deterministic* hidden variables. The volume did not actually appear until 1965, and so this paper is sometimes referenced as 1965. (I,A)

54. **"Introduction to the Hidden-Variable Question,"** J. S. Bell, in Ref. 1, pp. 171–181 (1971). A derivation of the inequality is given in Sec. 4, (entitled "A difficulty," one of the great understatements of the cen-

tury!). The generalization from *deterministic* to *stochastic* hidden variables is made here, but so briefly that it is easily overlooked. (I,A)

55. **"Atomic-cascade Photons and Quantum-mechanical Nonlocality,"** J. S. Bell, Com. At. Molec. Phys. **9**, 121–6 (1980). A clear and entertaining discussion of Bell's theorem, the experimental results, and their implications. (E)

3. From Gedankenexperiment to real experiments

56. **"Proposed Experiment to test Local Hidden-Variable Theories,"** J. F. Clauser, M. A. Horne, A. Shimony, and R. A. Holt, Phys. Rev. Lett. **23**, 880–4 (1969). (A)

*57. **"Experimental Test of Local Hidden-Variable Theories,"** S. J. Freedman and J. F. Clauser, Phys. Rev. Lett. **28**, 938–941 (1972). The experiment involves the correlation between the polarizations of two photons emitted in a $J = 0 \rightarrow J = 1 \rightarrow J = 0$ atomic cascade. (A)

58. **"Experimental Investigation of a Polarization Correlation Anomaly,"** J. F. Clauser, Phys. Rev. Lett. **36**, 1223–6 (1976). Repeat of an experiment (by Holt and Pipkin) that had contradicted QM; the new results agree with QM. (A)

59. **"Experimental Test of Local Hidden-Variable Theories,"** E. S. Fry and R. C. Thompson, Phys. Rev. Lett. **37**, 465–8 (1976). (A)

60. **"Proposed Molecular Test of Local Hidden-Variables Theories,"** T. K. Lo and A. Shimony, Phys. Rev. A **23**, 3003–12 (1981). Analysis of the feasibility of measuring the spin correlations of a pair of Na atoms from the decay of an unstable singlet excited state of Na_2. (A)

61. **"Experimental Tests of Realistic Local Theories via Bell's Theorem,"** A. Aspect, P. Grangier, and G. Roger, Phys. Rev. Lett. **47**, 460–3 (1981). (A)

62. **"Experimental Realization of Einstein–Podolsky–Rosen–Bohm *Gedankenexperiment*: A New Violation of Bell's Inequalities,"** A. Aspect, P. Grangier, and G. Roger, Phys. Rev. Lett. **49**, 91–4 (1982). (A)

*63. **"Experimental Test Bell's Inequalities Using Time-Varying Analyzers,"** A. Aspect, J. Dalibard, and G. Roger, Phys. Rev. Lett. **49**, 1804–7 (1982). The above three papers report a sequence of experiments on polarization correlations in pairs of atomic cascade photons, of increasing accuracy and sophistication, all of which confirm QM and contradict the Bell inequality. In the final experiment the polarizer directions are effectively switched between two orientations in a time shorter than the transit time for a light signal between them. (A)

64. **"Experimental Tests of Bell's Inequalities,"** A. Aspect, pp. 377–390, in **The Wave-Particle Dualism**, edited by S. Diner, D. Fargue, G. Lochak, and F. Selleri (Reidel, Dordrecht, 1984). A review of the experiments (Refs. 61, 62, and 63) and their implications. (I)

4. Simplifying the derivation

65. **"On Hidden Variables and Quantum Mechanical Probabilities,"** E. P. Wigner, Am. J. Phys. **38**, 1005–9 (1970). This is the first paper to relate Bell's theorem to the nonexistence of a non-negative joint probability distribution for the results of several experiments, not all of which can be performed together. (E)

*66. **"Cryptographic Approach to Hidden Variables,"** N. Herbert, Am. J. Phys. **43**, 315–6 (1975). A very simple derivation of a simplified version of Bell's inequality. (E)

*67. **"Upperformed Experiments have no Results,"** A. Peres, Am. J. Phys. **46**, 745–7 (1978). (E)

68. **"The Quantum Theory and Reality,"** B. d'Espagnat, Sci. Am. **241**, (5), 158–181 (Nov. 1979). An elementary account of Bell's inequality and its experimental tests. It is marred by an abstract stating, "The doctrine that the world is made up of objects whose existence is independent of human consciousness turns out to be in conflict with quantum mechanics and with facts established by experiment." That statement is devoid of any foundation because no assumption about the role of consciousness is made in the derivation of Bell's inequality. (E)

69. **"Bringing Home the Atomic World: Quantum Mysteries for Anyone,"** N. D. Mermin, Am. J. Phys. **49**, 940–3 (1981). (E)

70. **"Is the Moon there when Nobody looks? Reality and the Quantum Theory,"** N. D. Mermin, Phys. Today **38**, (4) 38–47 (1985). (E)

71. **"A Lecture Demonstration of the Incompatibility of Quantum Predictions with those of a Local Realistic Theory,"** G. E. Stedman, Am. J. Phys. **53**, 1143–9 (1985). A logic circuit is made to light bulbs in accordance with the QM predictions, in violation of any local theory. The device is essentially a realization of the imaginary device described by Mermin (Ref. 70). (E)

5. Strengthening and generalizing the inequality

*72. **"Experimental Consequences of Objective Local Theories,"** J. F. Clauser and M. A. Horne, Phys. Rev. D **10**, 526–535 (1974). This analysis treats stochastic, as well as deterministic, hidden-variable models, and treats in detail the limitations imposed by detector and polarizer efficiencies. (A)

73. **"On the Consequences of Einstein Locality,"** F. Selleri, Found. Phys. **8**, 103–116 (1978). Systematic derivation of several Bell-type inequalities. (A)

74. **"Generalized Inequalities Following from Einstein Locality,"** A. Garuccio, Lett. Nuovo Cimento **23**, 559–565 (1978). (A)

75. **"Systematic Derivation of all the Inequalities of Einstein Locality,"** A. Garuccio and F. Selleri, Found. Phys. **10**, 209–216 (1980). (A)

76. **"Joint Distributions and Local Realism in the Higher-Spin Einstein–Podolsky–Rosen Experiment,"** N. D. Mermin and G. M. Schwartz, Found Phys. **12**, 101–135 (1982). (A)

*77. **"Bell Inequalities with a Range of Violation that does not Diminish as the Spin Becomes Arbitrarily Large,"** A. Garg and N. D. Mermin, Phys. Rev. Lett. **49**, 901–4 (1982). (A) Erratum: Phys. Rev. Lett. **49**, 1294 (1982).

78. **"Local Realism and Measured Correlations in the Spin-s Einstein–Podolsky–Rosen Experiment,"** A. Garg and N. D. Mermin, Phys. Rev. D **27**, 339–348 (1983). (A)

79. **"Some Evaluations of Bell's Inequality for Particles of Arbitrary Spin,"** M. Ogren, Phys. Rev. D **27**, 1766–1773 (1983). Some results for higher spin are derived by a different method from that used by Mermin *et al.* (I)

80. **"Farkas's Lemma and the Nature of Reality: Statistical Implications of Quantum Correlations,"** A. Garg and N. D. Mermin, Found. Phys. **14**, 1–39 (1984). A general method is given for determining whether a set of pair distributions are consistent with the existence of higher-order distributions, of which they are marginal. The method is applied to quantum spin correlations, yielding generalized Bell-type inequalities. (A)

81. **"Generalizations of Bell's Theorem to Higher Spins and Higher Correlations,"** N. D. Mermin, in Ref. 5, pp. 7–20 (1986). Derivation of Bell-type inequalities that are violated by QM for every finite value of spin. A discussion is given of how these violations may disappear in the classical ($s \rightarrow \infty$) limit. (I)

6. Loopholes, real and imaginary

82. **"Hidden-Variable Example Based upon Data Rejection,"** P. M. Pearle, Phys. Rev. D **2**, 1418–25 (1970). A deterministic HV model is constructed in which some fraction of the particles go undetected. On the assumption that the experimenter will reject events in which only one (or zero) particles is detected, the remaining coincidence data can agree with the predictions of QM. (A)

83. **"On Hidden-Variable Theories and Bell's Inequality,"** L. de la Pena, A. M. Cetto, and T. Brody, Lett. Nuovo Cimento **5**, 177–181 (1972). It is claimed that the derivation of Bell's inequality is implicitly based on the postulate that "the measuring process has no discernable influence on the system subjected to it." This claim seems hard to sustain for the many careful derivations that have followed. (A)

84. **"Resolution of the Einstein–Podolsky–Rosen and Bell Paradoxes,"** I. Pitowsky, Phys. Rev. Lett. **48**, 1299 (1982). It is claimed that a local hidden-variable model which agrees with QM and violates the Bell inequalities can be constructed by means of nonmeasurable sets. The following two comments point out that contradictions between such models and QM can be derived without any assumption of measurability. (A)

85. **"Comment on 'Resolution ⋯,' "** N. D. Mermin, Phys. Rev. Lett. **49**, 1214 (1982).

86. **"Comment on 'Resolution ⋯,' "** A. L. Macdonald, Phys. Rev. Lett. **49**, 1215 (1982).

87. **"Pitowsky Responds,"** Phys. Rev. Lett. **49**, 1216 (1982).
88. **"Local Realism has not been Refuted by Atomic Cascade Experiments,"** T. W. Marshall, E. Santos, and F. Selleri, Phys. Lett. **98A**, 5–9 (1983). A local hidden-variable model is constructed which fits experimental data as well as does the prediction of QM. However, it is only the low efficiency of the present detectors that makes this possible (see Ref. 96). More accurate measurements could discriminate between such local models and QM. (A)
89. **"Bell's Theorem: Does the Caluser–Horne Inequality Hold for All Local Theories?,"** T. D. Angelidis, Phys. Rev. Lett. **51**, 1819–22 (1983). The negative answer given to the question in the title is based upon an error, as was pointed out by several authors.
90. **"Comment on 'Bell's Theorem: Does ···?,"** A. Garg and A. J. Leggett, Phys. Rev. Lett. **53**, 1019–20 (1984). (E)
91. **"Rotational Invariance, Locality, and Einstein–Podolsky–Rosen Experiments,"** A. O. Barut and P. Meystre, Phys. Rev. Lett. **53**, 1021 (1984). (E)
92. **"Angelidis Responds,"** Phys. Rev. **53**, 1022–3 (1984).
93. **"Comment on 'Bell's Theorem: Does ··?,"** M. A. Horne and A. Shimony, Phys. Rev. Lett. **53**, 1296 (1984). (E)
94. **"Angelidis Responds,"** Phys. Rev. **53**, 1297 (1984).
95. **"Comment on Angelidis's Universality Claim,"** J. T. Cushing, Phys. Rev. Lett. **54**, 2059 (1985). (E)
96. **"The EPR Experiment—Thoughts about the 'Loophole,' "** N. D. Mermin, in Ref. 7. Although the correlations predicted by QM definitely violate the Bell inequalities, it is possible to exploit the low efficiency (15%–20%) of current photon detectors to devise models such that the "real" correlations would obey the Bell inequalities (locality) but the observed data would appear to agree with QM. Papers 82, 72, and 88 contain examples of such models. This paper shows that the loophole for such models can be closed if and only if the detector efficiency exceeds 83%. (A)

7. Interpretation and significance of the results

97. **"Comments on Stapp's 'Copenhagen Interpretation' and the Significance of Bell's Theorem,"** L. E. Ballentine, Am. J. Phys. **42**, 81–3 (1974). Argues that since the assumptions of "local causality" which leads to Bell's inequality is violated in experiments, we should give up the "local" restriction, rather than "causality"; also argues that it is incorrect to claim that the experimental contradiction of Bell's inequality supports the Copenhagen interpretation. (A)
98. **"Reply to Ballentine's Comments,"** H. P. Stapp, Am. J. Phys. **42**, 83–5 (1974). (A)
99. **"Bell's Theorem without Hidden Variables,"** P. H. Eberhard, Nuovo Cimento **38B**, 75–9 (1977). A Bell inequality is derived without using the concepts of hidden variables or determinism. (A)
100. **"Bell's Theorem and the Different Concepts of Locality,"** P. H. Eberhard, Nuovo, Cimento **46B**, 392–418 (1978). The logical relations among four different formalized notions of locality are discussed. Part of the results of Ref. 110 are anticipated in this paper. (A)
101. **"Determinism and Locality in Quantum Mechanics,"** I. Nordin, Synthese **42**, 71–90 (1979). Argues that Bell's inequalities can be derived without any assumption of determinism, and that the observed violation of them indicates an essential nonlocality of QM and does not support any particular interpretation of QM. (A)
102. **"Hidden Variables, Joint probability, and the Bell Inequalities,"** A. Fine, Phys. Rev. Lett. **48**, 291–5 (1982). Claims a far-reaching equivalence among the concepts mentioned in the title. The results seems to be correct up to a point, but the scope of their validity is disputed in the following comment and response. (A)
103. **"Comment on 'Hidden Variables, Joint probability, and the Bell Inequalities,' "** A. Garg and N. D. Mermin, Phys. Rev. Lett. **49**, 242 (1982).
104. **"Fine Responds,"** Phys. Rev. Lett. **49**, 243 (1982). (A)
105. **"Correlation Inequalities and Hidden Variables,"** A. Garg and N. D. Mermin, Phys. Rev. Lett. **49**, 1220–3 (1982). The controversy with Fine is settled with a counterexample to some of his claims. (A)
106. **"Bell's Theorem as a Nonlocality Property of Quantum Theory,"** H. P. Stapp, Phys. Rev. Lett. **49**, 1470–4 (1982). This paper and the following two seek to correct and/or clarify some of the points raised in Ref. 102. (A)
107. **"Constraints of Determinism and of Bell's Inequalities are not Equivalent,"** P. H. Eberhard, Phys. Rev. Lett. **49**, 1474–7 (1982). (A)
108. **"Comments on the Significance of Bell's Theorem,"** A. Fine, Phys. Rev. Lett. **49**, 1536 (1982). (A)
109. **"Nonseparability and the Tentative Descriptions of Reality,"** B. d'Espagnat, Phys. Rep. **110**, 201–264 (1984). An extended discussion of the subject and its ramifications. The author's philosophical interpretations are not entirely covincing, however the appendix, "Some disproved objections to the Bell theorem," is quite useful. (A)
110. **"On the Physical Significance of the Locality Conditions in the Bell Arguments,"** J. P. Jarrett, Nous **18**, 569–589 (1984). The *strong locality* condition that is used to derive the Bell inequalities is shown to be equivalent to the conjunction of *relativistic locality* and another condition which has to do with the "*completeness*" of the state descriptions of the theory. It is only the latter that fails in QM. (A)

*111. **"An Exchange on Local Beables,"** J. S. Bell; A. Shimony, M. A. Horne, and J. F. Clauser, Dialectica **39**, 85–110 (1985), reprinted from Epistemological Letters: no. 9 (1976), no. 13 (1976), no. 15 (1977), no. 18 (1978). (The hideous neologism in the title should be spelled "be-able" to indicate its pronunciation and etymology. It is intended to refer to that which exists, in contradistinction to that which can be measured, an "observable.") Bell attempts to formulate a notion of "local causality" that is independent of and weaker than determinism. He concludes that QM is not compatible with local causality. SHC object that Bell's conclusion does not follow without further assumptions. Bell concedes this point, emphasizing in response that the experimenters' ability to freely choose the settings of their instruments must be assumed, and that if those choices were predetermined by a suitable conspiracy the proof would not be valid. (A)

112. **"Bell's Theorem and the Foundations of Quantum Physics,"** H. P. Stapp, Am. J. Phys. **53**, 306–317 (1985). The author derives a Bell-type inequality that is violated by QM, and carefully argues that no assumption of *realism* or *determinism* is involved, therefore, the contradiction is strictly between *locality* and QM. (See Ref. 17, Sec. 3.8, for an account of a controversy about the possibility of eliminating all traces of determinism from the proof.) (A)
113. **"Comments on 'Locality, Bell's Theorem, and Quantum Mechanics,' "** H. P. Stapp, Found. Phys. **15**, 973–6 (1985). A distinction is made between two ideas of locality: (1) No signal can travel faster than light; (2) The course of events in one space–time region can be *in no way disturbed* by what is done in a space–time region that is spatially separated from the first. (1) is involved in relativity, whereas (2) was used by EPR and Bell. (A)
114. **"EPR: What has it taught us?"** H. P. Stapp, in Ref. 6, pp. 637–652 (1985). A discussion of the significance for future work of the concepts introduced by EPR (Ref. 50). (I,A)
115. **"Bell's Theorem: Does Quantum Mechanics Contradict Relativity?,"** L. E. Ballentine and J. P. Jarrett, Am. J. Phys. **55**, 696 (1987). Expanding on the ideas introduced in Ref. 110, the authors answer the question in the title negatively. *Strong locality* (needed to derive the Bell inequalities) is equivalent to *relativistic locality* plus *predictive completeness* of the state description. Violation of the latter condition by QM is responsible for the violation of the Bell inequalities. It has the additional consequence that QM is indeed *incomplete* in the EPR sense, notwithstanding any failure of the EPR locality condition. (I,A)

F. Experiments (other than for Bell inequalities)

116. **"On the Statistical Aspect of Electron Interference Phenomena,"** P. G. Merli, G. F. Missiroli, and G. Pozzi, Am. J. Phys. **44**, 306–7 (1976). A demonstration experiment shows the buildup of an electron diffraction pattern from the pointlike flashes of individual electrons. The developing pattern can be shown in the classroom on a TV screen. (E)
117. **"Coherence Effects in Neutron Diffraction and Gravity Experiments,"** D. M. Greenberger and A. W. Overhauser, Rev. Mod. Phys. **51**, 43–78 (1979). A detailed theoretical analysis of the neutron interferometer and the effects of gravity and acceleration. (A)
118. **"Gravity and Inertia in Quantum Mechanics,"** J.-L. Staudenmann, S. A. Werner, R. Colella, and A. W. Overhauser, Phys. Rev. A **21**,

1419–38 (1980). Senstive neutron interferometer experiments are able to measure the effects of gravity and the Earth's rotation of the phase of the neutron wavefunction. (I)

119. **"Search for a Nonlinear Variant of the Schrödinger Equation by Neutron Interferometry,"** C. G. Shull, D. K. Atwood, J. Arthur, and M. A. Horne, Phys. Rev. Lett. **44**, 765–8 (1980). A plausible nonlinear variant of the Schrödinger equation, which preserves Galilean invariance and the usual probability interpretation but possesses stable nonspreading wave-packet solutions, is subjected to experimental test. An upper bound of 3.4×10^{-13} eV is obtained for the magnitude of the nonlinear term. (I)

120. **"Neutron Optical Tests of Nonlinear Wave Mechanics,"** R. Gahler, A. G. Klein, and A. Zeilinger, Phys. Rev. A **23**, 1611–7 (1981). The upper bound for the nonlinear term is reduced to 3.3×10^{-15} eV by means of precise Fresnel diffraction measurements. (I)

*121. **"Explicit Experimental Verification of Quantum Spin-State Superposition,"** J. Summhammer, G. Badurek, H. Rauch, and U. Kischko, Phys. Lett. **90A**, 110–112 (1982). The single-crystal interferometer is use to split a polarized neutron beam into two parts, which separate by a few centimeters before being recombined. The initial spin state is polarized in the $+z$ direction. A spin flipper is inserted into one of the separated beams, so that upon recombination one has a superposition of $|+\rangle$ and $|-\rangle$ spin states, which is mathematically equivalent to a spin state polarized in the xy plane. This prediction of the superposition principle is confirmed. (I)

122. **"The Neutron Interferometer as a Device for illustrating the Strange Behavior of Quantum Systems,"** D. M. Greenberger, Rev. Mod. Phys. **55**, 875–905 (1983). An interesting account of how many *Gedankenexperiments* are now capable of realization by means of the neutron interferometer. In an appendix he analyzes the weak equivalence principle (the motion of a particle in a gravitational field is independent of the mass of the particle) in the classical limit, and shows why the applicability of the Schrödinger equation to a particle in a gravitational potential should not have been regarded as a forgone conclusion before its experimental confirmation (Ref. 118). (I)

123. **"Time-Dependent Superposition of Spinors,"** G. Badurek, H. Rauch, and J. Summhammer, Phys. Rev. Lett. **51**, 1015–18 (1983). An experiment similar to Ref. 121, except that a radio-frequency spin flipper is used, instead of a static spin flipper. This causes the energies (frequencies) of the partial beams to be different. Upon recombination of the beams one has a nonstationary state, and oscillations of the spin polarization at the beat frequency is observed. (I)

124. **"A Real 'Thought' Experiment for the Hydrogen Atom,"** I. E. McCarthy and E. Weigold, Am. J. Phys. **51**, 152–5 (1983). The momentum probability distribution for the electron on a hydrogen atom is measured and found to agree with the square of the momentum representation wavefunction. (I)

125 **"Experimental Confirmation of the Aharonov–Bohm Effect by Electron Holography,"** A. Tomomura, in Ref. 5, pp. 169–176 (1986). Very striking confirmation of the relative phase shift of two electron beams that have passed through field-free regions separated by a region of magnetic flux. The nonquantization of flux is also clearly demonstrated. (I)

126. **"Neutron Interferometric Double-Resonance Experiment,"** G. Badurek, H. Rauch, and D. Tuppinger, Phys. Rev. A **34**, 2600–8 (1986). An experiment similar to Ref. 123, except that rf spin flippers are placed in both beams of the interferometer. Coherence of the beams persists in spite of the spin-flip and energy transfer to every neutron. (I)

127. **"Diffraction of Atoms by Light: The Near-Resonance Kapitza–Dirac Effect,"** P. L. Gould, G. A. Ruff, and D. E. Princhard, Phys. Rev. Lett. **56**, 827–330 (1986). The diffraction of an atomic beam by an electromagnetic standing wave, predicted by Kapitza and Dirac in 1933, has been observed. (A)

G. Quantization of the electromagnetic field

128. **"The Concept of the Photon,"** M. O. Scully and M. Sargent, Phys. Today **25** (3), 38–47 (1972). Many phenomena (including the photoelectric effect and much of quantum optics) that are commonly thought to require the photon concept for their explanation, can in fact be explained by semiclassical methods that apply QM only to matter but not to radiation. Quantum electrodynamics is needed to explain certain field fluctuation phenomena, but the "fuzzy ball" picture of the photon often leads to unnecssary confusion. (I)

Further correspondence on this subject is contained in the Aug., Dec. 1972, and Feb. 1973 issues.

129. **"Experimental Limitations to the Validity of Semiclassical Radiation Theories,"** J. F. Clauser, Phys. Rev. A **6**, 49–54 (1972). It is pointed out that some polarization correlation measurements by Kosher and Commins in 1967 are able to exclude theories which treat the electromagnetic field classically and use QM only for matter. (A)

*130. **"Experimental Distinction between the Quantum and Classical Field-Theoretic Predictions for the Photoelectric Effect,"** J. F. Clauser, Phys. Rev. D **9**, 853–860 (1974). A demonstration that a photon, unlike a wave packet, cannot be split by a half-silvered mirror. (I)

131. **"Non-Classical Effects in the Statistical Properties of Light,"** R. Loudon, Rep. Prog. Phys. **43**, 913–949 (1980). The classical and quantum theories of second-order coherence are reviewed. Three inequalities are derived which are satisfied by the classical theory but may be violated by the quantum theory. Experimental violations of two of these inequalities are described. (A)

132. **"Violations of Bell's Inequality in Cooperative States,"** P. D. Drummond, Phys. Rev. Lett. **50**, 1407–1410 (1983). Bell's inequality is applied to a cooperative state (analogous to the superfluorescent state) of N photon pairs. [The experiments of Aspect *et al.* (Refs. 61–63) correspond to $N=1$.] As N becomes large the low-order correlation functions of the field cease to violate Bell's inequality, but violations always exist in sufficiently high-order correlations. (A)

133. **"Evidence for the Quantum Nature of Light,"** D. F. Walls, Nature **280**, 451–4 (1979). The "antibunching" of photons in resonance fluorescence is one of the few known phenomena which cannot be explained by any classical field theory of radiation. (I)

134. **"Nonclassical Radiation of a Single Stored Ion,"** F. Diedrich and H. Walther, Phys. Rev. Lett. **58**, 203–6 (1987). Radiation from a single atom exhibits antibunching. (I)

135. **"Photons in Introductory Quantum Physics,"** J. Strnad, Am. J. Phys. **54**, 650–2 (1986). Analogies between photons and electrons can be misleading. A photon wavefunction cannot be introduced except under very limited circumstances. (I)

H. Miscellaneous papers

136. **"Test of the Non-Separability of the $K^0\overline{K}^0$ System,"** J. Six, pp. 391–6, in *Quantum Mechanics a Half Century Later*, edited by J. L. Lopes and M. Paty (Reidel, Dordrecht, 1977). Proposes a test for EPR-type correlations in the decay of the $K^0\overline{K}^0$ system. (A)

137 **"Theory of Reality,"** H. P. Stapp, Found. Phys. **7**, 313–323 (1977). "Bell's theorem is used to guide the formulation of a unified theory of reality that incorporates the basic principles of relativistic quantum theory," (quoted from author's abstract). (A)

138. **"Locality and Reality,"** H. P. Stapp, Found Phys. **10**, 767–795 (1980). An attempt to formulate an interpretation of QM that is compatible with the ideas of both Bohr and Einstein, and with Bell's theorem. (A)

139. **"Probability Theory in Quantum Mechanics,"** L. E. Ballentine, Am. J. Phys. **54**, 883–9 (1986). It is demonstrated that the formalism of QM satisfies the axioms of probability theory. Some erroneous claims that QM violates classical probability theory are examined and refuted. (E)

140. **"Microphysical Reality,"** W. Duch and D. Aerts, Phys. Today **39**, (6), 13–5 (1986). A humorous poll of physicists regarding their beliefs about locality and QM. (E)

The Statistical Interpretation of Quantum Mechanics

L. E. BALLENTINE
Department of Physics, Simon Fraser University, Burnaby, B.C., Canada

The Statistical Interpretation of quantum theory is formulated for the purpose of providing a sound interpretation using a minimum of assumptions. Several arguments are advanced in favor of considering the quantum state description to apply only to an ensemble of similarily prepared systems, rather than supposing, as is often done, that it exhaustively represents an individual physical system. Most of the problems associated with the quantum theory of measurement are artifacts of the attempt to maintain the latter interpretation. The introduction of hidden variables to determine the outcome of individual events is fully compatible with the statistical predictions of quantum theory. However, a theorem due to Bell seems to require that any such hidden-variable theory which reproduces all of quantum mechanics exactly (i.e., not merely in some limiting case) must possess a rather pathological character with respect to correlated, but spacially separated, systems.

CONTENTS

INTRODUCTION

1.0 Preface and Outline

This article is not a historical review of how the quantum theory and its statistical interpretation came to be. That task has been admirably carried out by Max Jammer (1966) in his book *The Conceptual Development of Quantum Mechanics,* and also by van der Waerden (1967). Our point of departure can conveniently be introduced by considering the following statement made by Peierls (1967) in a review of Jammer's book:

"Chapter 7 ··· is headed 'The Copenhagen Interpretation.' ··· the phrase suggests that this is only one of several conceivable interpretations of the same theory, whereas most physicists are today convinced that the uncertainty relations and the ideas of complementarity are essential parts of the structure of quantum mechanics ···. A discussion of alternative interpretive attempts is promised for a later volume. ··· [But] even the cautious word 'attempts' may be too positive a description for what are only programs ··· that are not yet clear even in outline."

If, as appears to be the case, the latter remarks by Peierls refer to the models known as hidden-variable theories (see Sec. 6), we agree that these should be treated as new theories, and that they are not new interpretations of quantum mechanics "any more than quantum mechanics is a new interpretation of classical physics." However we shall show, contrary to the view expressed by Peierls, that the Copenhagen interpretation contains assumptions which are not "essential parts of the structure of quantum mechanics," and that one such assumption is at the root of most of the controversy surrounding "the interpretation of quantum mechanics." It is the assumption that the quantum state description is the most complete possible description of an individual physical system.

An interpretation which is more nearly minimal in the sense of including all verifiable predictions of quantum theory, but without the contestable features of the Copenhagen interpretation, we shall call the *Statistical Interpretation.* The distinction between these interpretations (which share many features in common) will be made in the following sections. Suffice it to say, for now, that if we identify the Copenhagen Interpretation with the opinions of Bohr, then the Statistical Interpretation is rather like those of Einstein. Contrary to what seems to be a widespread misunderstanding, Einstein's interpretation corresponds very closely with the one which is almost universally used by physicists in practice; the additional assumptions of the Copenhagen interpretation playing no real role in the applications of quantum theory.

The outline of this paper is as follows. First we give a brief summary of the mathematical formalism of quantum theory in order to distinguish the formalism, which we accept, from the physical interpretation, which we shall examine critically. We then consider different interpretations, and expound the Statistical Interpretation in detail.

The Secs. 2, 3, and 4 could be prefaced by Appendix *xi of Popper (1959) on the proper use of imaginary experiments, in which he points out that Gedanken experiments can be used to *criticize* a theory but not to *justify* or *prove* a theory. Our discussions of the Gedanken experiment of Einstein, Podolsky, and Rosen, of the uncertainty principle, and of the measurement process are undertaken to criticize the assumption that a state vector provides a complete description of an individual system. Although arguments of this type can refute the hypothesis being criticized, they cannot, of course, "prove" the Statistical Interpretation but can only illustrate its advantages.

Sections 5 and 6 deal with two concepts (joint probability distributions for position and momentum, and hidden variables) which have often been thought to be incompatible with quantum theory. That belief, however, was based in an essential way upon the above hypothesis which we criticize and reject. It turns out that, within certain limits, the formalism of quantum theory can be extended (*not* modified) to include these concepts within the Statistical Interpretation.

Finally we summarize our conclusions.

1.1 Mathematical Formalism of Quantum Theory

Quantum theory, and indeed any theory, can be divided [see Prugovecki (1967), or Tisza (1963)] into:

(a) A mathematical formalism consisting of a set of primitive concepts, relations between these concepts (either postulated or obtainable by given rules of deduction), and a dynamical law.

(b) Correspondence rules which relate the theoretical concepts of (a) to the world of experience.

This division is not absolute—clearly one must have a formalism in order to make correspondence rules, but unless one has at least some partial idea of correspondence rules, one would not know what one was talking about while constructing the formalism—nevertheless it is convenient for the present task.

The mathematical formalism of quantum theory is well known and can be abstracted from any of several textbooks (Dirac, 1958; Messiah, 1964). The primitive concepts are those of *state* and of *observable*.

F1 An *observable* is represented by a self-adjoint operator on a Hilbert space. It has a spectral representation,

$$R=\sum_n r_n P_n, \tag{1.1}$$

where the P_n are orthogonal projection operators related to the orthonormal eigenvectors of R by

$$P_n=\sum_a |\, a, r_n\rangle\langle a, r_n \,|. \tag{1.2}$$

Here the numbers r_n are the eigenvalues of R, and the parameter a labels the degenerate eigenvectors which belong to the same eigenvalue of R. The sums become integrals in the case of continuous spectra. Equation (1.1) is equivalent to the statement that an observable must possess a complete orthogonal set of eigenvectors.

F2 A *state* is represented by a state operator (also called a statistical operator or density matrix) which must be self-adjoint, nonnegative definite, and of unit trace. This implies that any state operator may be diagonalized in terms of its eigenvalues and eigenvectors,

$$\rho=\sum_n \rho_n \,|\, \phi_n\rangle\langle\phi_n \,|, \tag{1.3a}$$

with

$$0\leq\rho_n\leq 1 \tag{1.3b}$$

and

$$\sum_n \rho_n=1. \tag{1.3c}$$

This state operator formalism is reviewed by Fano (1957).

F3 A *pure state* can be defined by the condition $\rho^2=\rho$. It follows that for a pure state there is exactly one nonzero eigenvalue of ρ, say,

$$\rho_n=1, \qquad \rho_{n'}=0 \quad \text{for} \quad n\neq n'. \tag{1.4}$$

In this case we have

$$\rho= |\, \phi_n\rangle\langle\phi_n \,|, \tag{1.5}$$

and so a pure state may be represented by a vector in the Hilbert space. A general state which is not pure is commonly called a *mixed state*.

F4 The average value of an observable R in the state ρ is given by

$$\langle R\rangle= \mathrm{Tr}\,(\rho R), \tag{1.6}$$

where Tr means the trace of the operator in parentheses. For a pure state represented by the normalized vector $|\psi\rangle$, (1.6) reduces to $\langle R\rangle=\langle\psi\,|\,R\,|\,\psi\rangle$. By introducing the characteristic function $\langle e^{i\xi\cdot R}\rangle$, we can obtain the entire statistical distribution of the observable R in the state ρ. It follows that:

F5 The only values which an observable may take on are its eigenvalues, and the probabilities of each of the eigenvalues can be calculated. In the case of a pure state represented by the normalized vector $|\psi\rangle$, the probability of eigenvalue r_n of R is $\sum |\langle\psi\,|\,a, r_n\rangle|^2$. This is a generalization of Born's (1926) famous postulate that the square of a wave function represents a probability density.

So far we have only given necessary, but not sufficient, conditions for the mathematical representations of observables and of states. To complete the specification, the following is usually postulated:

F6 The Hilbert space is a direct sum of coherent

subspaces, within each of which (almost) every vector may represent a pure state. This is a formal statement of the *superposition principle* with allowance being made for superselection rules.[1] The set of all mixed states can be constructed from the set of all pure states using (1.3).

F7 Any self-adjoint operator which commutes with the generators of superselection rules, or equivalently, all of whose eigenvectors lie within coherent subspaces of F6, represents an observable. This postulate may be criticized on the grounds that it is difficult to imagine a procedure for observing a quantity like $(x^2p_x^3+p_x^3x^2)$, which should be an observable according to this postulate. The general form of F7 is unnecessary in most, if not all practical applications, but it is used in Von Neumann's theorem (Sec. 6.1).

F8 The dynamical law or equation of motion depends in detail upon the physical system under consideration (i.e., number of degrees of freedom, whether relativistic or nonrelativistic), but in every case it can be written in the form,

$$\rho(t)=U\rho(t_0)U^{-1} \tag{1.7a}$$

in general, or

$$|\psi(t)\rangle=U|\psi(t_0)\rangle \tag{1.7b}$$

for a pure state, where $U=U(t, t_0)$ is a unitary operator.

The above is not intended to be an axiomatization of quantum theory, but merely a compact summary of the mathematical formalism of the theory as it exists at present and in practice. Except for the reservation noted in F7 it should be noncontroversial. Such is not the case with the correspondence rules.

1.2 Correspondence Rules

The correspondence rules must relate the primitive concepts of *state* and *observable* to empirical reality. In so doing they will provide a more specific interpretation for the averages and probabilities introduced in F4 and F5.

The natural requirement placed upon an *observable* is that we should be able to observe it. More precisely, an *observable* is a dynamical variable whose value can, in principle, be measured. For canonically conjugate variables the corresponding operators are obtained through Dirac's canonical commutation relation,

$$qp-pq=\hbar i. \tag{1.8}$$

There is no general rule for constructing a unique operator to represent an arbitrary function $f(q, p)$ because of the noncommutability of q and p (Shewell, 1959). Fortunately the most general case does not seem to arise in practice. We can, for example, unambiguously determine the Hamiltonian operator for any finite number of particles interacting through velocity-independent potentials in the presence of an arbitrary external electromagnetic field.

The different interpretations of quantum theory are most sharply distinguished by their interpretations of the concept of *state*. Although there are many shades of interpretation (Bunge, 1956), we wish to distinguish only two:

(I) The *Statistical Interpretation*, according to which a *pure state* (and hence also a general state) provides a description of certain statistical properties of an *ensemble* of similarily prepared systems, but need not provide a complete description of an individual system.

This interpretation is upheld, for example, by Einstein (1949), by Popper (1967), and by Blokhintsev (1968). Throughout this paper the capitalized name "Statistical Interpretation" refers to this specific interpretation (described in detail in Sec. 1.3).

(II) Interpretations which assert that *a pure state provides a complete and exhaustive description of an individual system* (e.g., an electron).

This class contains a great variety of members, from Schrödinger's original attempt to identify the electron with a wave packet solution of his equation to the several versions of the Copenhagen Interpretation.[2] Indeed many physicists implicitly make assumption II without apparently being aware that it is an additional assumption with peculiar consequences. It is a major aim of this paper to point out that the hypothesis II is unnecessary for quantum theory, and moreover that it leads to serious difficulties.

1.3 The Statistical Interpretation

The term, Statistical Interpretation, will be used throughout this paper in the specific sense here expounded, and should not be confused with the less specific usage of this term by other writers (e.g., Messiah, 1964, Chap. IV) who do not distinguish it from the Copenhagen interpretation.

Of primary importance is the assertion that a quantum

[1] A superselection rule (Wick *et al.*, 1952; Galindo *et al.*, 1962) is a restriction on the superposition principle. For example, a vector which is a linear combination of integer and half-integer angular momentum eigenvectors cannot represent a physical state. A *coherent* space is one in which the superposition principle has unrestricted validity.

[2] Bohr (1935, 1949) argued against Einstein, who rejected II. Heisenberg's position is somewhat unclear. One reference (1930, p. 33) contains a statement fully in accord with the Statistical Interpretation; however, in a later writing specifically in defense of the Copenhagen interpretation (1955, p. 26) he states "an individual atomic system can be represented by a wave function . . . ". His interpretation of probability (1958, Chap. 3) was based on the Aristotelian notion of "potentia", which is quite different from the Statistical Interpretation. Messiah (1964) is quite explicit (p. 152, 158) in favoring II over I. Although both claim orthodoxy, there now seems to be a difference of opinion between what may be called the Copenhagen school represented by Rosenfeld, and the Princeton school represented by Wigner (see Rosenfeld, 1968 and references therein). But since both factions appear to accept hypothesis II, our criticism will apply to both.

state (pure or otherwise) represents an *ensemble* of similarly prepared systems. For example, the system may be a single electron. Then the ensemble will be the conceptual (infinite) set of all single electrons which have been subjected to some state preparation technique (to be specified for each state), generally by interaction with a suitable apparatus. Thus a momentum eigenstate (plane wave in configuration space) represents the ensemble whose members are single electrons each having the same momentum, but distributed uniformly over all positions. A more realistic example which occurs in scattering problems is a finite wave train with an approximately well-defined wavelength. It represents the ensemble of single electrons which result from the following schematically described procedure—acceleration in a machine, the output from which can take place in only some finite time interval (due to a "chopper"), and collimation which rejects any particle whose momentum is outside certain limits. We see that a quantum *state* is a mathematical representation of the result of a certain state preparation procedure. Physical systems which have been subjected to the same state preparation will be similar in some of their properties, but not in all of them (similar in momentum but not position in the first example). Indeed the physical implication of the uncertainty principle (discussed in detail in Sec. 3) is that no state preparation procedure is possible which would yield an ensemble of systems identical in all of their observable properties. Thus it is most natural to assert that a quantum state represents an ensemble of similarily prepared systems, but does not provide a complete description of an individual system.

When the physical system is a single particle, as in the above examples, one must not confuse the *ensemble*, which is a conceptual set of replicas of one particle in its experimental surroundings, with a *beam* of particles, which is another kind of (many-particle) system. A beam may simulate an ensemble of single-particle systems if the intensity of the beam is so low that only one particle is present at a time.

The ensembles contemplated here are different in principle from those used in statistical thermodynamics, where we employ a representative ensemble for calculations, but the result of a calculation may be compared with a measurement on a single system. Also there is some arbitrariness in the choice of a representative ensemble (microcanonical, canonical, grand canonical). But, in general, quantum theory predicts nothing which is relevant to a single measurement (excluding strict conservation laws like those of charge, energy, or momentum), and the result of a calculation pertains directly to an ensemble of similar measurements. For example, a single scattering experiment consists in shooting a single particle at a target and measuring its angle of scatter. Quantum theory does not deal with such an experiment, but rather with the statistical distribution (the differential cross section) of the results of an ensemble of similar experiments. Because this ensemble is not merely a representative or calculational device, but rather it can and must be realized experimentally, it does not inject into quantum theory the same conceptual problems posed in statistical thermodynamics.

In general quantum theory will not predict the result of a measurement of some observable R. But the probability of each possible result r_n, calculated according to F5, may be verified by repeating the state preparation and the measurement many times, and then constructing the statistical distribution of the results. As pointed out by Popper (see Körner, 1957, p. 65, p. 88), one should distinguish between the *probability*, which is the relative frequency (or measure) of the various eigenvalues of the observable in the conceptual infinite ensemble of all possible outcomes of identical experiments (the sample space), and the *statistical frequency* of results in an actual sequence of experiments. The probabilities are properties of the state preparation method and are logically independent of the subsequent measurement, although the statistical frequencies of a long sequence of similar measurements (each preceded by state preparation) may be expected to approximate the probability distribution. If hypothesis I is adopted, we may say simply that the probabilities are properties of the *state*.

The various interpretations of a quantum *state* are related to differences in the interpretation of probability (see Popper, 1967 for a good exposition of this point). In contrast to the Statistical Interpretation, some mathematicians and physicists regard probability as a measure of knowledge, and assert that the use of probability is necessitated only by the incompleteness of one's knowledge. This interpretation can legitimately be applied to games like bridge or poker, where one's best strategy will indeed be influenced by any knowledge (accidentally or illegally obtained) about an opponent's cards. But physics is not such a game, and as Popper has emphasized, one cannot logically deduce new and verifiable knowledge—statistical knowledge—literally from a lack of knowledge.

Heisenberg (1958), Chap. 3, combined this "subjective" interpretation with the Aristotelian notion of "potentia." He considers a particle to be "potentially present" over all regions for which the wave function $\psi(\mathbf{r})$ is nonzero, in some "intermediate kind of reality," until an act of observation induces a "transition from the possible to the actual." In contrast, the Statistical Interpretation considers a particle to always be at some position in space, each position being realized with relative frequency $|\psi(\mathbf{r})|^2$ in an ensemble of similarily prepared experiments. The "subjective" and Aristotelian ideas are primarily responsible for the suggestion that the observer plays a peculiar and essential role in quantum theory. Whether or not they can be consistently developed, the existence of the Statistical Interpretation demonstrates that they are *not necessary*, and in my opinion they bring with them no advantages to compensate for the additional metaphysical complication.

If the expression "wave-particle duality" is to be used at all, it must not be interpreted literally. In the above-mentioned scattering experiment, the scattered portion of the wave function may be equally distributed in all directions (as for an isotropic scatterer), but any one particle will not spread itself isotropically; rather it will be scattered in some particular direction. Clearly the wave function describes not a single scattered particle but an ensemble of similarily accelerated and scattered particles. At this point the reader may wonder whether a statistical particle theory can account for interference or diffraction phenomena. But there is no difficulty. As in any scattering experiment, quantum theory predicts the statistical frequencies of the various angles through which a particle may be scattered. For a crystal or diffraction grating there is only a discrete set of possible scattering angles because momentum transfer to and from a periodic object is quantized by a multiple of $\Delta p = h/d$, where Δp is the component of momentum transfer parallel to the direction of the periodic displacement d. This result, which is obvious from a solution of the problem in momentum representation, was first discovered by Duane (1923), although this early paper had been much neglected until its revival by Landé (1955, 1965). There is no need to assume that an electron spreads itself, wavelike, over a large region of space in order to explain diffraction scattering. Rather it is the crystal which is spread out, and the electron interacts with the crystal as a whole through the laws of quantum mechanics. For a longer discussion of this and related problems such as the two-slit experiment, see Landé (1965).[3] In every case a diffraction pattern consists of a statistical distribution of discrete particle events which are separately observable if one looks in fine enough detail. In the words of Mott (1964, p. 409), "Students should not be taught to doubt that electrons, protons and the like are particles ··· The wave cannot be observed in any way than by observing particles."

Although we shall discuss his work in greater detail below, we should emphasize here the great contribution of Einstein to this subject. His "Reply to Criticisms" (Einstein, 1949), expressed very clearly his reasons for accepting a purely statistical (ensemble) interpretation of quantum theory, and rejecting the assumption that the state vector provided an exhaustive description of the individual physical system.

2. THE THEOREM OF EINSTEIN, PODOLSKY, AND ROSEN

The tenability of hypothesis II, Sec. 1.2, was challenged in a paper by Einstein, Podolsky, and Rosen (1935) (abbreviated EPR) entitled "Can Quantum-Mechanical Description of Reality Be Considered Complete?" Their argument is often referred to as the Paradox of EPR, as though it ought to be capable of resolution as, say, Zeno's paradox. [Rosenfeld (1968) has scurrilously referred to it as the EPR fallacy.] We shall show however that, properly interpreted, it is a well defined theorem, paradoxical only to the extent that the reader may not have expected the conclusion.

In order to precisely answer the question posed in their title, EPR introduce the following definitions:

D1 A *necessary* condition for a *complete* theory is that *"every element of physical reality must have a counterpart in the physical theory."*

D2 A *sufficient* condition for identifying an *element of reality* is, *"If, without in any way disturbing a system, we can predict with certainty* (i.e., *with probability equal to unity*)[4] *the value of a physical quantity, then there exists an element of physical reality corresponding to this physical quantity."*

Their argument then proceeds by showing, through consideration of a thought experiment, that two non-commuting observables should, under suitable conditions, both be considered elements of reality. Since no state vector can provide the value of both of these observables, they conclude that the quantum state vector cannot completely describe an individual system, but only an ensemble of similarily prepared systems.

2.1 A Thought Experiment and the Theorem

The experiment described below was introduced by Bohm (1951, p. 614ff). By considering measurements of spins rather than of particle positions and momenta (as in the original EPR experiment), we avoid any unnecessary complications with the position-momentum uncertainty principle which may arise from the minimum degree of particle trajectory definition needed to perform an experiment.

Two particles of spin one-half are prepared in an unstable initial state of total spin zero. The pair separates, conserving total spin, and one of the particles (which are chosen to be distinguishable, for convenience) passes through the inhomogeneous magnetic field of a Stern-Gerlach apparatus (see Fig. 1).

---

[3] I note in passing that Landé's ambitious program to derive all of quantum theory from a few simple principles has not yet been completely successful. Reviewers of his book (Shimony, 1966; Witten, 1966) have pointed out additional assumptions implicit in his argument which are virtually equivalent to assuming some of the results he wishes to derive. However, these criticisms do not detract from his discussion of diffraction scattering.

[4] As a nearly pedantic refinement, we would prefer to replace this phrase by *with probability* $1-\epsilon$, *where* ϵ *may be made arbitrarily small*. This allows for arbitrarily accurate approximations to problems which may not be formally solvable, and it allows one to avoid certain irrelevant criticisms. (See Footnote 5).

The uncertainty principle is trivially satisfied since we require only that $p_x = p_z = 0$ approximately, and that Δy be small enough so we know which particle has entered the magnetic field.

The translational motion of the particles can be treated classically, and the spin can be described by a spin Hamiltonian in which the magnetic field at the position of the particle is treated as a function of time,

$$H = V_{12} + V_H. \tag{2.1}$$

Here $V_{12} = V(\mathbf{r}_{12}, \boldsymbol{\sigma}_1, \boldsymbol{\sigma}_2)$ is the interaction between the two particles, and V_H is the interaction with the magnetic field.

$$\begin{aligned} V_H(t) &= zH'\sigma_{1z}, && \text{for} \quad 0<t<\tau, \\ &= 0, && \text{for} \quad t<0 \text{ or } t>\tau, \end{aligned} \tag{2.2}$$

where τ is the transit time of the particle through the magnetic field which is of gradient H' and is directed along the z axis.

Omitting the position-dependent factors, we can write the initial state vector, a spin singlet, as

$$\psi(0) = \{u_+(1)u_-(2) - u_-(1)u_+(2)\}/\surd 2. \tag{2.3}$$

Here u_+ and u_- are the two-component spinors corresponding to the eigenvalues ± 1 of σ_z, and the arguments 1 or 2 refer to the two particles. Since

$$\psi(t) = \exp\left\{\left(-\frac{i}{\hbar}\right)\int_0^t H(t')\,dt'\right\}\psi(0), \tag{2.4}$$

the effect of the interparticle interaction V_{12} will be negligible if

$$|V_{12}|\,\tau/\hbar \ll 1, \tag{2.5}$$

a condition which can be realized by making the transit time τ short enough.[5] By $|V_{12}|$ we mean the modulus of the relevant matrix elements of V_{12} in (2.4).

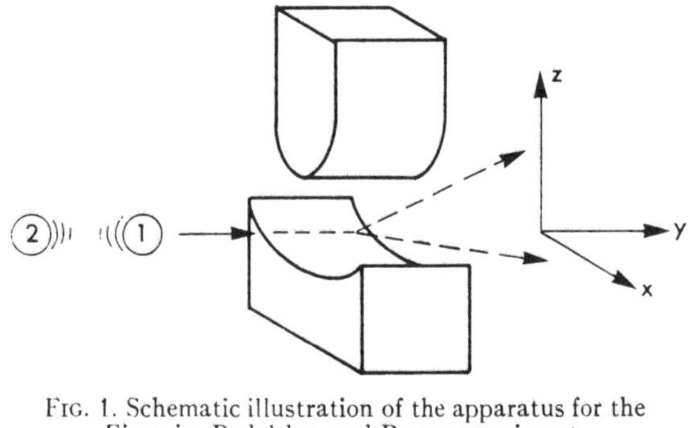

FIG. 1. Schematic illustration of the apparatus for the Einstein, Podolsky, and Rosen experiment.

[5] This overcomes the objection of Sharp (1961) that any interaction (even gravitational) would prevent us from writing $\psi(t)$ as a product of two factors; one for particle 1, and one for particle 2. Such an objection is quite irrelevant because any state vector can be expressed as a linear combination of product-type basis vectors, and the interaction only affects the time dependence of the expansion coefficients which we treat to an arbitrary degree of precision.

At time $t=\tau$, after the particle has interacted with the magnetic field, the state vector will be

$$\psi(\tau) = \{\exp(-iH'\tau z_1/\hbar)u_+(1)u_-(2) - \exp(iH'\tau z_1/\hbar)u_-(1)u_+(2)\}/\surd 2. \tag{2.6}$$

The result of the interaction is to produce a correlation between z components of momentum and spin of particle 1, and spin of particle 2. If $p_{1z} = -H'\tau$, then $\sigma_{1z} = +1$, and $\sigma_{2z} = -1$; or if $p_{1z} = H'\tau$ then $\sigma_{1z} = -1$ and $\sigma_{2z} = +1$. It is only necessary to make the magnetic field gradient H' large enough so that the two values of p_{1z} are unambiguously separated.

We have thus shown that the z component of the spin of particle 2 can be determined to an arbitrarily high degree of accuracy by a measurement which, because of the spacial separation and negligible effect of V_{12}, does not in any way disturb particle 2. Thus according to definition D2, σ_{2z} is an *element of reality.*

However the initial singlet state is invariant under rotation, and can equally be expressed as

$$\psi(0) = \{v_+(1)v_-(2) - v_-(1)v_+(2)\}/\surd 2, \tag{2.7}$$

where the spinors v_+ and v_- are eigenvectors of σ_x. If the Stern–Gerlach magnet were rotated so that the field was directed along the x axis, then by an identical argument we would conclude that σ_{2x} was an *element of reality.* Since no state vector can provide a value for both of the noncommuting observables σ_{2x} and σ_{2z}, EPR conclude, in accordance with D1, that the state vector does not provide a complete description of an individual system.

One might try to avoid this conclusion by adopting an extreme positivist philosophy, denying the *reality* of both σ_{1z} and σ_{2z} until the measurement has actually been performed. But this entails the unreasonable, essentially solipsist position that the reality of particle 2 depends upon some measurement which is not connected to it by any physical interaction.

In any case, the conclusion can be stated, as was first done by Einstein (1949, p. 682), as the following *theorem*:

The following two statements are incompatable:

(1) The state vector provides a complete and exhaustive description of an individual system;

(2) The real physical conditions of spatially separated (noninteracting) objects are independent.

Of course one is logically free to accept either one of these statements (or neither). Einstein clearly accepted the second, while Bohr apparently favored the first. The importance of the EPR argument is that it proved for the first time that assuming the first statement above *demands* rejection of the second, and vice versa, a fact that was not at all obvious before 1935, and which may not be universally realized today.

2.2 Discussion of the EPR Theorem

It should be emphasized that the result of the EPR experiment is in no way paradoxical. The implication of (2.6) in the Statistical Interpretation is that if the experiment is repeated many times we should obtain the result $p_{1z}=-H'\tau$, $\sigma_{1z}=+1$, $\sigma_{2z}=-1$ in about one-half of the cases, and the opposite result in the other half of the cases. This correlation between the spins of the two particles will be the same no matter which components are measured.

Another essential point, which has not always been realized, is that the EPR theorem in no way contradicts the mathematical formalism of quantum theory. It was only intended to exhibit the difficulties which follow from the assumption that the state vector completely describes an individual system; an interpretive assumption which is not an integral part of the theory. In his famous reply to EPR, Bohr (1935) said, "Such an argumentation [as that of EPR], however, would hardly seem suited to affect the soundness of quantum-mechanical description, which is based on a coherent mathematical formalism ⋯." It would appear that Bohr failed to distinguish between the mathematical formalism of quantum theory, and the Copenhagen interpretation of that theory. Einstein's criticism was directed only at the latter, not at the former. Conversely the self-consistency and empirical success of the former are no defense against specific criticism of the latter. Such an erroneous perception of the EPR paper as being an attack on quantum theory itself may explain (psychologically) the often repeated statement that Bohr had successfully refuted their argument. In fact Bohr's paper offers no real challenge to the validity of the EPR theorem as stated above, nor does the EPR theorem pose any "paradox" or threat to the validity of quantum theory.

Bohr's reply to EPR is really a criticism of their definitions of *completeness* and *physical reality*, but it is of a rather imprecise character. A satisfactory pursuit of this line of attack should first admit the validity of the EPR theorem with their definitions. Second it should propose alternative definitions with arguments in favor of the superiority of the new definitions. Finally it should show that the state vector provides a *complete* description of *physical reality* in terms of the new definitions. Bohr has done none of these.

In the succeeding years numerous comments on the EPR paper have been published, many of which are less than satisfactory. A discussion of a paper by Furry (1936) is given in Footnote 13, Sec. 4.3 of this paper. Breitenberger (1965) has criticized many authors for contributing to confusion, including Bohr for suggesting that the observation of particle 1 "creates" the physically real state of particle 2—a position approaching the absurdity of solipsism.[6] Breitenberger correctly emphasizes that there is nothing paradoxical about the EPR experiment, which is a prototype for many coincidence experiments, but he seems to forget the original purpose of the EPR argument.

Several people have proposed experiments similar in principle to that of EPR (Day, 1961; Inglis, 1961; Bohm and Aharonov, 1957). While these experiments are interesting in their own right, it should be emphasized that their results will not distinguish between Einstein's and Bohr's interpretations of quantum theory. If the results of experiments were to differ from the theoretical predictions, this would contradict the formalism of quantum theory itself, not just one of the interpretations.

3. THE UNCERTAINTY PRINCIPLE

3.1 Derivation

To a given *state* there will, in general, correspond a statistical distribution of values for each *observable*. A suitable measure of the width of the distribution for an *observable* A is the variance,

$$(\Delta A)^2=\langle(A-\langle A\rangle)^2\rangle, \tag{3.1}$$

where ΔA is known as the standard deviation of the distribution. Although *states* exist (at least as mathematical idealizations) for which the variance of the distribution for any one observable is arbitrarily small, it can easily be shown, as was first done by Robertson (1929) and Schrödinger (1930), that the product of the variances of the distributions of two observables A and B has a lower bound,

$$(\Delta A)^2(\Delta B)^2\geq(\tfrac{1}{2}\langle AB+BA\rangle-\langle A\rangle\langle B\rangle)^2+\tfrac{1}{4}\langle C\rangle^2, \tag{3.2}$$

where $AB-BA=iC$. The first term may vanish, but for canonically conjugate observables [see Eq. (1.8)] we must always have

$$\Delta q\Delta p\geq\hbar/2. \tag{3.3}$$

The meaning of these results is unambiguous. The averages of quantum theory (postulate F4) are realized by performing the same measurement on many similarly prepared systems (or equivalently by performing the measurement many times on the same system which must be resubmitted to the same state preparation before each measurement). In order to measure ΔA (or ΔB) one must measure A (or B) on many similarily prepared systems, construct the statistical distribution of the results, and determine its standard deviation. The results (3.2) and (3.3) assert that for any particular *state* (i.e., state preparation) the product of the widths of the distributions of A measurements and of B measurements may not be less than some lower limit. A term such as *the statistical dispersion principle* would really be more appropriate for these results than the traditional name, *uncertainty principle*. A discussion similar to this is given by Margenau (1963).

[6] I have not found an explicit statement to this effect in Bohr's writings, although essentially this position has been taken by some followers of Bohr.

3.2 Relation to Experiments

There exist many widespread statements of the uncertainty principle which are different from the one that is derived from statistical quantum theory. Very common is the statement that *one cannot measure the two quantities q and p simultaneously without errors whose product is at least as large as* $\hbar/2$. This statement is often supported by one or both of the following arguments:

(i) A measurement of q causes an unpredictable and uncontrollable disturbance of p, and vice versa. [This was first proposed by Heisenberg (1927) and is widely repeated in text books].

(ii) The position and momentum of a particle do not even exist with simultaneously and perfectly well defined (though perhaps unknown) values (Bohm, 1951, p. 100).

Clearly the statistical dispersion principle and the common statement of the uncertainty principle are not equivalent or even closely related. The latter refers to *errors* of *simultaneous* measurements of q and p on *one* system, and it is plausible that one of these measurements could cause an error in the other. On the other hand, the former refers to *statistical spreads* in *ensembles* of measurements on similarily prepared systems. But only one quantity (either q or p) is measured on any one system, so there is no question of one measurement interfering with the other. Furthermore the standard deviations Δq and Δp of the statistical distributions cannot be determined unless the errors of the individual measurements, δq and δp are much smaller than the standard deviations (see Fig. 2). These points have been emphasized by Margenau (1963) and by Popper (1967). Prugovecki (1967) has pointed out that, far from restricting simultaneous measurements of noncommuting observables, quantum theory does not deal with them at all; its formalism being capable only of statistically predicting the results of measurements of one observable (or a commutative set of observables). In order to deal with simultaneous measurements he proposes an extension of the mathematical formalism, to which we shall return in Sec. 5 of this paper.

We now consider to what extent the common statement of the uncertainty principle may be true, even

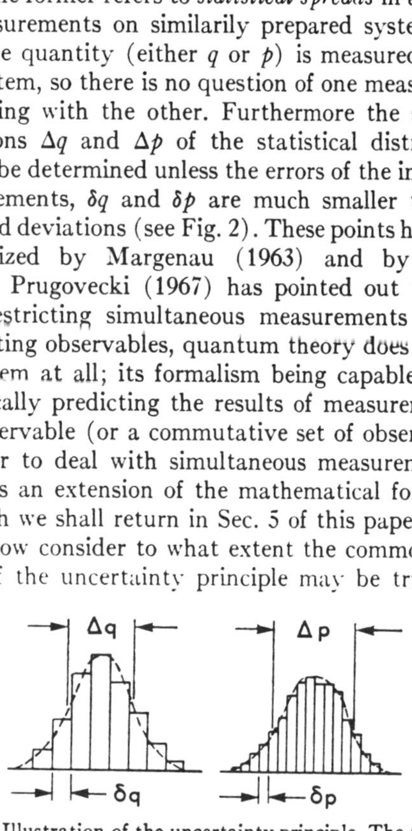

FIG. 2. Illustration of the uncertainty principle. The two histograms represent the frequency distributions of independent measurements of q and p on similarly prepared systems. There are, in principle, no restrictions on the precisions of individual measurements δq and δp, but the standard deviations will always satisfy $\Delta q \Delta p \geq \hbar/2$.

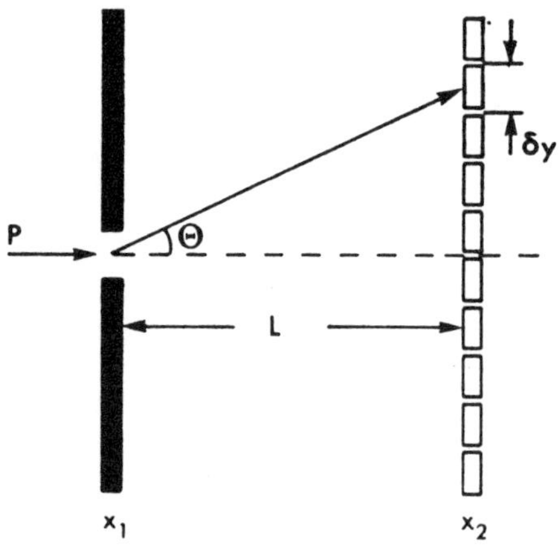

FIG. 3. An experimental arrangement designed to simultaneously measure y and p_y, such that the error product $\delta y \delta p_y$ can be arbitrarily small.

though it is not derivable from quantum theory. Argument (ii) is easily seen to be unjustified. It is based on the obvious fact that a wave function with a well defined wavelength must have a large spacial extension, and conversely a wave function which is localized in a small region of space must be a Fourier synthesis of components with a wide range of wavelengths. Using de Broglie's relation between momentum and wavelength, $p=h/\lambda$, it is then asserted that a particle cannot have definite values of both position and momentum at any instant. But this conclusion rests on the almost literal identification of the particle with the wave packet (or what amounts to the same thing, the assumption that the wave function provides an exhaustive description of the properties of the particle). The untenable nature of such an identification is shown by the example of a particle incident upon a semitransparent mirror with detectors on either side. The particle will either be reflected or transmitted without loss of energy, whereas the wave packet is divided, half its amplitude being transmitted and half reflected. A consistent application of the Statistical Interpretation yields the correct conclusion that the division of the wavepacket yields the relative probabilities for transmission and reflection of particles. But there is no justification for assertion (ii).

Argument (i) must be considered more seriously since Heisenberg (1930) and Bohr (1949) have given several examples for which it is apparently true. However, Fig. 3 shows a simple experiment for which it is not valid. A particle with known initial momentum p passes through a narrow slit in a rigid massive screen. After passing through the hole, the momentum of the particle will be changed due to diffraction effects, but its energy will remain unchanged. When the particle strikes one of the distant detectors, its y coordinate is thereby measured with an error δy. Simultaneously this same event serves to measure the y component of momentum, $p_y = p \sin\theta$, with an error δp_y which may be made arbitrarily small by making the distance L

arbitrarily large. Clearly the product of the errors δp_y need not have any lower bound, and so the common statement of the uncertainty principle given above cannot be literally true. One may raise the objection that p_y has not been measured, but only defined in the above equation. However this method of measuring momentum by means of geometrical inference from a position measurement is universally employed in scattering experiments. It rests upon the assumption of linear motion in a field-free region (Newton's First Law), which remains valid in quantum mechanics (at least for L much greater than a de Broglie wavelength).

The distinction between experiments which satisfy and which violate the uncertainty principle can be clarified with the help of the concepts of *state preparation* and *measurement*; the distinction between these having been emphasized by Margenau (1958, 1963) and by Progovecki (1967).

State preparation refers to any procedure which will yield a statistically reproducible ensemble of systems. The concept of *state* in quantum theory (see Sec. 1.3) can be considered operationally as an abbreviation for a description of the state preparation procedure.[7] Of course there may be more than one experimental procedure which yields the same statistical ensemble, i.e., the same state. An important special case (which is sometimes incorrectly identified with measurement) is a filtering operation, which ensures that if a system passes through the filter it must immediately afterward have a value of some particular observable within a restricted range of its eigenvalue spectrum.

On the other hand, *measurement* of some quantity R for an individual system means an interaction between the system and a suitable apparatus, so that we may infer the value of R (within some finite limits of accuracy) which the system had immediately before the interaction (or the value of R which the system would have had if it had not interacted, allowing for the possibility that the interaction will disturb the system).

The essential distinctions between the two concepts are that state preparation refers to the future, whereas measurement refers to the past; and equally important, that measurement involves *detection* of a particular system, whereas state preparation provides conditional information about a system *if* it passes through the apparatus.

The *statistical dispersion principle* (3.2, 3.3), which follows from the formalism of quantum theory, is a statement about the minimum dispersion possible in any state preparation. It is significant that experiments which satisfy the uncertainty principle can be employed as state preparations, whereas experiments which violate the uncertainty principle cannot.

Consider, for example, Heisenberg's (1930, p. 21) famous microscope for measuring the position of an electron. If the angular aperture of the microscope is ϵ and the wavelength of light used is λ, then the accuracy of the position measurement will be limited to

$$\delta x=\lambda/\sin\epsilon, \tag{3.4}$$

by the resolving power of the instrument. (We use the symbol δ to indicate the uncertainty or error in an individual measurement, while Δ refers to the standard deviation of an ensemble of similar measurements). Because the direction of the scattered photon is unknown within a cone of angle ϵ, the x component of momentum of the electron will be changed by some unknown increment in the range $\pm\sin\epsilon(h/\lambda)$. Note that this experiment is *not* an example of simultaneous measurement of x and p_x. Only x is measured here.

Suppose now that we have a state preparation apparatus which will produce an ensemble of electrons with a negligible spread in momentum (i.e., $\Delta p_x\approx 0$). Suppose that we try to select from this ensemble a subensemble which has as small as possible a spread Δx. We do this by measuring the x coordinate of each particle and selecting those which fall within the minimum resolvable range δx of a particular value, say, x'. Thus our subensemble will have a spread

$$\Delta x=\lambda/\sin\epsilon. \tag{3.5}$$

Because the recoil momentum absorbed by an electron may vary, the spread in momentum among the members of this subensemble will be

$$\Delta p_x\approx\sin\epsilon(h/\lambda). \tag{3.6}$$

Hence the ensemble which we have selected will have statistical dispersion of magnitude $\Delta x\Delta p_x\approx h$.

Contrast the above result with the experiment illustrated in Fig. 3. In this case we are able to determine the values of the position and momentum of a particle to an accuracy $\delta y\delta p_y\ll\hbar$. But this information refers to the motion of the particle during a certain interval *before* the measurement. We cannot use it to generate an ensemble of particles with statistical spreads $\Delta y\Delta p_y\ll\hbar$ because the scattering of the particles by the detector will cause a spread of momentum in the final ensemble, $\Delta p_y\gg\delta p_y$. If we try to avoid this scatter by removing one of the detectors from the array at the plane $x=x_2$, and inferring from a negative response of all the remaining detectors that the particle passed through the hole, then the momentum spread of the ensemble formed in this way will not be δp_y (the error of measurement described above). Rather it will be $\Delta p_y\approx\delta y/\hbar$ (due to diffraction effects). The spread of particle positions in the ensemble will be $\Delta y\approx\delta y$ (the size of the hole), and so the *statistical dispersion principle* will be satisfied.

In conclusion, the uncertainty principle restricts the degree of statistical homogeneity which it is possible to achieve in an ensemble of similarly prepared systems,

[7] Lamb (1969) has described an idealized method of preparing an arbitrary single-particle state.

and thus it limits the precision which future predictions for any system can be made. But it does not impose any restriction on the accuracy to which an event can be reconstructed from the data of both state preparation and measurement in the time interval between these two operations. Heisenberg (1930, p. 20) made nearly this distinction in his statement, "··· the uncertainty relation does not refer to the past." His subsequent remark, "It is a matter of personal belief whether ··· the past history of the electron [as inferred from an experiment like that of Fig. 3] can be ascribed any physical reality or not,"—based on the fact that it cannot be reconfirmed by any other future measurement, seems unduely cautious. In fact, the majority of real physical measurement are of just this type. Indeed for every future oriented experimental operation which yields verifiable information, there must be another past oriented operation whose function is not to be verifiable but to verify (Popper, 1967, pp. 25–28). It is just this essential fact which is emphasized by the distinction between *state preparation* and *measurement*.

3.3 Angular and Energy–Time Relations

An apparent contradiction of the uncertainty principle, which is frequently rediscovered by bright students, concerns the polar angle φ and the z component of angular momentum L_z. If L_z is represented by $-i\hbar\partial/\partial\varphi$, then it would appear that $[\varphi, L_z]=i\hbar$, from which it would follow that

$$\Delta\varphi\Delta L_z\geq\hbar/2 \quad \text{(wrong)}. \tag{3.7}$$

But (3.7) is obviously wrong since ΔL_z can be arbitrarily small, and φ has a meaningful range of only 2π radians; hence, $\Delta\varphi\leq 2\pi$.

The resolution of the apparent paradox, as pointed out by Judge and Lewis (1963) and by Susskind and Glogower (1964), lies in the observation that the operator $-i\hbar(\partial/\partial\varphi)$ is Hermitian only on the space of functions of φ which have period 2π, and multiplication by φ destroys this periodicity property. In order to deduce a correct uncertainty relation, one must replace the coordinate φ by some periodic function. This can be done in many ways but the most convenient seems to be the use of $\cos\varphi$ and $\sin\varphi$ (Carruthers and Nieto, 1968), from which it can be shown that

$$(\Delta L_z)^2[(\Delta\cos\varphi)^2+(\Delta\sin\varphi)^2] \geq(\hbar/2)^2(\langle\sin\varphi\rangle^2+\langle\cos\varphi\rangle^2). \tag{3.8}$$

The derivation of the energy-time uncertainty relation cannot follow the standard form (3.2) because time is not usually represented by an operator. In fact one can show (Susskind and Glogower, 1964) that, if the energy spectrum has a lower bound, then there does not exist a Hermitian operator which is canonically conjugate to the Hamiltonian in the sense of (1.8). However, the following result can be deduced from (3.2) (Messiah, 1964, p. 319),

$$\Delta A\Delta E\geq\tfrac{1}{2}\hbar\mid d\langle A\rangle/dt\mid, \tag{3.9}$$

where A is an arbitrary observable, and the averages are calculated for an arbitrary time-dependent state. If one defines a characteristic time for the system and the state as

$$\tau=\min_{(A)}\{\Delta A\mid d\langle A\rangle/dt\mid^{-1}\}, \tag{3.10}$$

then one may write

$$\tau\Delta E\geq\tfrac{1}{2}\hbar. \tag{3.11}$$

Clearly this result does not imply that one cannot measure the energy of a system exactly at an instant of time, as is sometimes stated, but rather that the spread of energies associated with any state is related to the characteristic rate of change associated with the same state.[8]

4. THE THEORY OF MEASUREMENT

The desirability of an analysis of the measurement process by means of quantum theory was perhaps first indicated through Bohr's insistence that the "*whole experimental arrangement*," (Bohr, 1949, p. 222), object plus all apparatus, must be taken into account in order to specify well-defined conditions for an application of the theory. However he never carried this program to its logical conclusion; the description of both the object and the measuring apparatus by the formalism of quantum theory. Such an analysis is useful for several reasons.

It has frequently been asserted that the acts of measurement or observation play a different role in quantum theory than in the rest of physics. Such claims should be critically analyzed.

Because the measurement apparatus usually (always?) contains a final stage to which classical mechanics is applicable, the measurement of a quantal (i.e., essentially nonclassical) object by such an apparatus involves the union of micro and macrophysics in an essential way. Any proposed revisions at the microscopic level, of the concept of physical reality or of the role of the observer will meet a critical test here. It is widely believed that Ehrenfest's theorem (Messiah, p. 216) proves the consistency of quantum theory with the classical limit, but this is only partly true. Ehrenfest's theorem demonstrates that the average values of observables obey the classical equations of motion provided the quantum state function is such that statistical deviations of the basic observables (coordinates and momenta) from their averages are negligible. But this leaves open the question of whether or not the state of a macroscopic object coupled to microscopic objects, as calculated from the quantal equation of motion for realistic but general initial

[8] After completion of this manuscript, a paper by Allcock (1969) containing a fuller discussion of energy-time relations appeared.

conditions, will necessarily possess this property. As will be shown, this appears not to be so in the case of measurement.

It is not the compatibility of the classical and quantal equations of motion (in the appropriate limit) which concerns us but rather the compatibility of the classical and quantal concepts of *state*. Thus an analysis of measurement will be very helpful in deciding the question raised in Sec. 1.2, that is whether a quantum *state* describes an individual system (object plus apparatus in this case), or whether it must refer only to an ensemble of systems.

4.1 Analysis of the Measurement Process

The essence of a measurement is an interaction between the object to be measured and a suitable apparatus so that a correspondence is set up between the initial state of the object and the final state of the apparatus. This interaction may or may not change the value of the observable being measured, the final state of the object being of no significance to the success of the measurement.

Suppose we wish to measure the observable R of the object I, for which there must be a complete set of eigenvectors,

$$R \mid \mathrm{I}; r\rangle = r \mid \mathrm{I}; r\rangle.$$

Denote a set of states for the apparatus II by $\mid \mathrm{II}; \alpha\rangle$, where the eigenvalue α is the appropriate "pointer position" of the apparatus, and $\mid \mathrm{II}; 0\rangle$ is the initial premeasurement state. The interaction between the object and the apparatus ideally should, if it is to function as a measurement, set up a unique correspondence between the initial value of r and the final value of α, α_r, for example. That is, if the initial state for the system I+II is $\mid \mathrm{I}; r\rangle \mid \mathrm{II}; 0\rangle$, then the equation of motion must lead to the final state

$$U \mid \mathrm{I}; r\rangle \mid \mathrm{II}; 0\rangle = \mid \mathrm{I}; r\rangle \mid \mathrm{II}; \alpha_r\rangle, \qquad (4.1a)$$

if the observable R is not changed by the interaction[9,10]; or in the general case

$$U \mid \mathrm{I}; r\rangle \mid \mathrm{II}; 0\rangle = \mid \mathrm{I}; \phi_r\rangle \mid \mathrm{II}; \alpha_r\rangle, \qquad (4.1b)$$

where U is the time-development operator for the duration of the interaction between I and II. Note that the vectors $\mid \mathrm{I}; \phi_r\rangle$ need not be orthogonal since orthogonality of the final states corresponding to different r values is guaranteed by the fact that $\alpha_{r'} \neq \alpha_r$, for $r' \neq r$.

Suppose now that the initial state of the object is not an eigenvector of R, but rather some linear combination

$$\mid \mathrm{I}; \psi\rangle = \sum_r \mid \mathrm{I}; r\rangle\langle r \mid \psi\rangle. \qquad (4.2)$$

From (4.1) and the linearity of the equation of motion, the final state for the system must be

$$U \mid \mathrm{I}; \psi\rangle \mid \mathrm{II}; 0\rangle = \sum_r \langle r \mid \psi\rangle \mid \mathrm{I}; \phi_r\rangle \mid \mathrm{II}; \alpha_r\rangle$$
$$= \mid \mathrm{I+II}; f\rangle, \quad \text{say}. \qquad (4.3)$$

A specific illustration of this general analysis is provided by the EPR experiment discussed in Sec. 2.1. We may consider the measured quantity R to be the spin of particle 2, and the "pointer" of the apparatus to be the momentum of particle 1 (assuming that two detectors are placed so as to determine its sign).

According to the Statistical Interpretation and F5 of the formalism, the probability of the apparatus "pointer" being α_r at the end of an experiment is $\mid \langle r \mid \psi\rangle \mid^2$. That is to say, if the experiment were repeated many times, always preceded by the same state preparation for both object and apparatus, the relative frequency of the value α_r would be $\mid \langle r \mid \psi\rangle \mid^2$. That this should be identical with the probability of the object having had the value r for the dynamical variable R immediately before interacting with the apparatus II, is just the criterion that II functions as a measuring apparatus for R.

4.2 The Difficulties of the "Orthodox" Interpretation

The discussion of the analysis of measurement according to the Statistical Interpretation was so simple and natural that further comment almost seems redundant. But if instead of the basic assumption of the Statistical Interpretation that a state vector characterizes an *ensemble* of similarily prepared systems, one assumes that a state provides a complete description of an *individual* system, then the situation is quite different. Unfortunately this assumption has been made so frequently, more often tacitly than not, that it and the resulting point of view have come to be considered "orthodox".

The "orthodox" quantum theory of measurement was formalized by J. Von Neumann (1955) and has been clearly reviewed by Wigner (1963). The first difficulty encountered is that if one tries to interpret the final-state vector (4.3) as completely describing one system (object plus apparatus), then one is led to say that the pointer of the apparatus has no definite position, since (4.3) is a coherent superposition of vectors corresponding to different values of α. But the pointer can be a quite classical object and it makes no sense to say that it has no position at any instant of time, especially since the different values of α_r are macro-

[9] Many authors give the unwarranted impression that this highly special case is general.

[10] Araki and Yanase (1960) have shown that unless R commutes with all universal additively conserved quantities, then no interaction exists which could satisfy (4.1a). However, they also show that this equation can nevertheless be satisfied to an arbitrary degree of accuracy, provided the apparatus is made arbitrarily large. This fact is illustrated in the measurement of the z component of spin by a Stern–Gerlach apparatus (Sec. 2.1), which used an external magnetic field to break the conservation of S_x and S_y. The source of this magnetic field is rigidly fixed to the earth, and so the apparatus is effectively infinite, as it must be to satisfy the theorem of Araki and Yanase. See also Yanase (1961).

scopically distinguishable. Thus if we attribute the state vector to an individual system, we are inevitably led to classically meaningless states for a classical object.[11]

To circumvent this difficulty, the "orthodox" theory assumes that, in addition to the continuous evolution of the state vector according to the equation of motion, there is also an unpredictable discontinuous "*reduction of the state vector*" upon "measurement," from (4.3) to

$$| \mathrm{I}; \phi_{r'} \rangle \, | \mathrm{II}; \alpha_{r'} \rangle, \tag{4.4}$$

where $\alpha_{r'}$ is the pointer position which is actually observed. Proponents of this point of view also usually assume that the value of the observable R is not changed by measurement, so that in place of (4.4) they would obtain

$$| \mathrm{I}; r' \rangle \, | \mathrm{II}; \alpha_{r'} \rangle, \tag{4.4$'$}$$

but this point is not essential. The word "measurement" is enclosed in quotes because its use in this peculiar context of the "orthodox" theory is not equivalent to the more physical definition of measurement given in Sec. 3.

It should be emphasized that this *reduction* cannot arise from the equation of motion. The form of (4.3), involving a superposition of macroscopically distinct pointer positions, depends only upon the linearity of the equation of motion and not upon any of our simplifying assumptions. For example, one might object that the pointer position eigenvalue α is not sufficient to label a unique state vector for the apparatus, since there are a huge number of commuting observables. But this means only that each vector $| \mathrm{II}; \alpha \rangle$ must be replaced by a set of vectors, all the sets being mutually orthogonal. Komar (1962) has shown that even with these extra degrees of freedom, a final state of the form (4.4) is not possible.[12] If the *reduction of the state vector* is to enter the theory at all, it must be introduced as a special postulate (often called the *projection postulate*).

If one merely applied the projection postulate to the object I, one would say that upon "measurement" the state changes discontinuously from (4.2) to $| \mathrm{I}; r' \rangle$, where r' is the "measured" value of the observable R. On the other hand, the analysis of the interaction between the object I and the apparatus II leads to the state (4.3). To avoid a direct contradiction, Von Neumann introduced the observer III, who supposedly performs a "measurement" on I+II (this "measurement" however is merely an observation), and thus *reduces* the state vector to (4.4) or (4.4′). He then shows that it makes no difference whether the apparatus II is considered to be part of the observer or part of the object being observed.

Von Neumann's theory of measurement is very unsatisfactory. It suggests that the passive act of observation by a conscious observer is essential to the understanding of quantum theory. Such a conclusion is without foundation, for we have seen that the Statistical Interpretation does not require any such considerations. Moreover, the atoms of the observer's body may, in principle, be treated by quantum theory, and the resulting state vector will be similar to (4.3) (with I replaced by I+II, and II replaced by III). Von Neumann's theory would require another hypothetical observer IV to observe III, and so *reduce* the state vector. But this can only lead to an infinite regression, which, in an earlier era might have been taken as a proof of the existence of God (the Ultimate Observer)! No one seems to have drawn this unfounded conclusion, but the equally unfounded conclusion that the consciousness of an observer should be essential to the theory has, unfortunately, been taken seriously (Heitler, 1949, p. 194; Wigner, 1962).

Because of the difficulties inherent in the projection postulate, we must critically examine the reasons given in support of its introduction. Dirac (1958, p. 36) argues that a second measurement of the same observable immediately after the first measurement must always yield the same result. Clearly this could be true at most for the very special class of measurements that do not change the quantity being measured. A statement of such limited applicability is hardly suitable to play any fundamental role in the foundations of quantum theory. In fact this argument is based on the implicit (and incorrect) assumption that *measurement* is equivalent to *state preparation*, the contrasting definitions of which were given in Sec. 3. The necessity for distinguishing between these two concepts has been pointed out by Margenau (1958; 1963). For example, a polaroid filter placed in the path of a photon beam constitutes a *state preparation* with respect to the polarization of any transmitted photons. A second polaroid at the same angle has no further effect. But neither of these processes constitutes a *measurement*. To measure the polarization of a photon one must also detect whether or not the photon was transmitted through the polaroid filter. Since the detector will absorb the photon, no second measurement is possible.

It has also been claimed (Messiah, 1964, p. 140) that a discontinuous *reduction* of the state vector is a consequence of an alleged uncontrollable perturbation of the object by the measuring apparatus. A counter example to this claim is provided by the EPR experiment (Sec. 2.1), in which the spin of particle 2 is

[11] This was first pointed out by Schrödinger (1935), who proposed the following hypothetical experiment. A cat is placed in a chamber together with a bottle of cyanide, a radioactive atom, and a device which will break the bottle when the atom decays. One half-life later the state vector of the system will be a superposition containing equal parts of the living and dead cat, but any time we look into the chamber we will see either a live cat or a dead one. A literal translation of "Schrödinger's cat paradox", as this is called, has been given by Jauch (1968, p. 185), but his discussion of the paradox is not adequate, and his volume and page reference is incorrect.

[12] His conclusion that no theory consistent with quantum mechanics can account for the occurrence of events in nature, however, is not true if one drops the assumption that a state vector describes completely an individual system.

measured indirectly, without any disturbance of that particle whatsoever. In general the apparatus will interact with the object, but the effect of this is not such as to bring about the hypothetical *reduction*. We do not consider as credible the occasional statement (Dirac, 1958, p. 110) that a mere observation (i.e., an experimenter looking at his apparatus) could affect the state of the system in a manner incompatible with the equation of motion.

Although it is not essential to the theory, Popper (1967) has pointed out that the *reduced* state vector (4.4) can be interpreted in a natural way. According to the Statistical Interpretation, the state vector (4.3) represents, not an individual system, but the ensemble of all possible systems (object plus apparatus), each of which has been prepared in a certain way and then allowed to interact. The state vector (4.4) represents the subensemble whose definition includes the additional specification that the result of the measurement (pointer reading) be $\alpha_{r'}$. There is no question of *reduction of the state vector* being a physical process.

Margenau (1963, p. 476) has taken a more sympathetic attitude toward Von Neumann's theory of measurement on the grounds (gathered from personal conversation) that he regarded his projection postulate as convenient but not necessary. In reply I would say that I find no evidence of this interpretation in Von Neumann's writings. Moreover, the great amount of confusion generated, as well as the theoretical effort expended trying to explain the *reduction of the state vector*, suggest that the projection postulate has been much more of an inconvenience than a convenience.

4.3 Other Approaches to the Problem of Measurement

An enormous number of papers have been written on the problem of measurement in quantum theory. Margenau (1963), lists over 60 separate articles, and promises a list of about 200 on request. The reader can shorten his task greatly by ignoring all papers which try, without modifying quantum theory, to accomodate a *reduction of the state vector*, and which also assume the state vector to describe an individual system. The preceding arguments have demonstrated the impossibility of such programs.

Bohm and Bub (1966) propose a nonlinear modification of the Schrödinger equation of motion which is supposed to be effective only during the measurement process, and which causes the *reduction*. However we do not consider that the postulation of a new interaction peculiar to measurement should be taken seriously. Moreover, the simplicity of the account of measurement provided by the Statistical Interpretation undermines the motivation for any drastic modification of the mathematical formalism.

To discuss the next approach it is convenient to restate the results of our analysis in the language of state operators (or statistical operators). After the interaction between the object I and the apparatus II, the final state operator for the combined system I+II will be the pure state

$$\rho_f = |\,\mathrm{I}+\mathrm{II}; f\rangle\langle \mathrm{I}+\mathrm{II}; f\,|, \qquad (4.5)$$

$$|\,\mathrm{I}+\mathrm{II}; f\rangle = \sum_r c_r \,|\,\mathrm{I}; \phi_r\rangle\,|\,\mathrm{II}; \alpha_r\rangle, \qquad (4.3')$$

where $c_r = \langle r \mid \psi\rangle$ in the notation of (4.3). Now the mixed state formed from *reduced* vectors of the type (4.4),

$$\rho_m = \sum_r |\,c_r\,|^2\,|\,\mathrm{I}; \phi_r\rangle\,|\,\mathrm{II}; \alpha_r\rangle\langle \mathrm{I}; \phi_r\,|\,\langle \mathrm{II}; \alpha_r\,| \qquad (4.6)$$

is not equivalent to ρ_f. Even though they both yield the same probability distribution for the position of the apparatus "pointer" α, their predictions for other observables may disagree.[13] Because time development is effected by a unitary transformation (1.7a), which preserves the pure state condition $\rho^2 = \rho$, it is impossible for the pure state (4.5) ever to evolve into the mixed state (4.6).

Realizing this fact, many authors (Feyerabend, 1957; Wakita, 1960, 1962; Daneri *et al.*, 1962; Jauch, 1964) have proposed that, nevertheless, (4.5) and (4.6) may be equivalent for all practical purposes. Specifically, they suggest that if the macroscopic nature of the apparatus and its relevant observables are taken into account, then the average of any *macroscopic* observable M will be the same for these two states. With the abbreviated notation $|\,\phi_r, \alpha\rangle \equiv |\,\mathrm{I}; \phi_r\rangle\,|\,\mathrm{II}; \alpha\rangle$, these averages are, respectively,

$$\mathrm{Tr}\,(\rho_f M) = \sum_{r,r'} c_{r'} c_r^* \langle \phi_r, \alpha_r \mid M \mid \phi_{r'}, \alpha_{r'}\rangle, \qquad (4.7)$$

$$\mathrm{Tr}\,(\rho_m M) = \sum_r |\,c_r\,|^2 \langle \phi_r, \alpha_r \mid M \mid \phi_r, \alpha_r\rangle. \qquad (4.8)$$

The problem is then whether these expressions are equal under sufficiently general circumstances.

Now if M commutes with A (the pointer position, whose eigenvalues are α_r), it will be diagonalizable in this representation, and the interference terms ($r \neq r'$) will be absent from (4.7). Hence the two expressions will be equal. However, (contrary to Jauch's assumption) the set of macroscopic observables is not Abelian (consider the pointer momentum, which does not commute with A), so the macroscopic nature of M is not sufficient to guarantee equality of (4.7) with (4.8).

In the case where the observable M belongs entirely

---

[13] A proof of this fact is contained in a paper by Furry (1936). But his comments on the work of Einstein, Podolsky, and Rosen (Sec. 2) are based upon a misconception. Furry interprets the EPR conclusion that certain dynamical variables such as σ_{2z} in our Eq. (2.6) are *elements of reality*, as meaning that the state of the system after interaction between the object and the apparatus is a mixture of eigenstates of the *element of reality*; i.e., a mixed state related to (2.6) in the same way as (4.6) is related to (4.5). But EPR do not claim that the equation of motion yields the wrong quantum state, as Furry has misinterpreted them, but rather that no quantum state, pure or mixed, can provide a complete description of an individual system. This makes Furry's argument irrelevant.

to the measurement apparatus II (i.e., if $M=1\otimes M_{\text{II}}$),[14] then

$$\langle\phi_r,\alpha_r\mid M\mid\phi_{r'},\alpha_{r'}\rangle = \langle\text{I};\phi_r\mid\text{I};\phi_{r'}\rangle\langle\text{II};\alpha_r\mid M_{\text{II}}\mid\text{II};\alpha_{r'}\rangle. \quad (4.9)$$

Thus if the states $\mid\text{I};\phi_r\rangle$ are mutually orthogonal, which would be true, in particular, if the measurement of R did not change the value of R, ($\mid\text{I};\phi_r\rangle$ then becomes $\mid\text{I};r\rangle$), then there would be no difference between (4.7) and (4.9). Clearly a similar result holds for observables belonging entirely to the object I, but of course no such result is valid for observables which belong to I and II jointly.[15]

Daneri, Loinger, and Prosperi (1962, 1966) (DLP) consider the pure state (4.5) or equivalently (4.3) to be the final state of the combined system *immediately* after interaction between the object and the apparatus. They then concentrate on the amplification aspect of the measurement necessary for a microscopic object to trigger a macroscopic response in the apparatus, and invoke the ergodic theory of quantum statistical mechanics in order to treat the transition of the apparatus from its initial "metastable"[16] condition to equilibrium. They claim, in essence, that the equilibrium state can be represented by the mixed state operator (4.6).

The attitude of DLP is well expressed in the words of Rosenfeld (1965, p. 230), who endorses their approach: "The reduction of the initial state of the atomic system has nothing to do with the interaction between this system and the measuring apparatus: in fact, it is related to a process taking place in the latter apparatus after all interaction with the atomic system has ceased." Rosenfeld also emphasizes the view that the presence or absence of a conscious human observer is irrelevant to the problem.[17]

The above remarks show that DLP and Rosenfeld explicitly reject the "orthodox" interpretation discussed in Sec. 4.2 (as does the present author), but they do so without actually confronting the question of whether the state operator (or vector) represents a single system or an ensemble of similarily prepared systems.

The analysis of generalized quantum amplification devices by DLP is interesting and correct, at least to the extent that their ergodic hypothesis is valid. But it does not constitute the essence of measurement in general. To paraphrase an example given by Jauch (1968, p. 169), if the measuring device is a photographic plate, then the essence of measurement may be performed by a single silver–halide complex. The amplification process does not take place until the plate is developed, which may be months later. Although interesting in its own right, the DLP theory does not answer the questions which lead us to study the theory of measurement.

4.4 Conclusion—Theory of Measurement

This section on measurement theory is lengthy because of the great length of the literature on this subject. However, the conclusions are quite simple. The formal analysis of a measurement has been undertaken not for its own sake, but only to test the consistency of alternative interpretations of quantum theory. Using only the linearity of the equation of motion and the definition of measurement, we see that the interaction between the object and the measuring apparatus leads, in general, to a quantum state which is a coherent superposition of macroscopically distinct "pointer positions." In the Statistical Interpretation this dispersion of pointer positions merely represents the frequency distribution of the possible measurement results for a given state preparation.

But if one assumes that the state vector completely describes an individual system, then the dispersion must somehow be a property of the individual system, but it is nonsensical to suppose that a macroscopic pointer has no definite position. None of the attempts to solve this problem using some form of *reduction of the state vector* are satisfactory. No such problem arises in the Statistical Interpretation, in which the state vector represents an ensemble of similarily prepared systems, so it must be considered superior to its rivals.

We have neglected the frankly subjective interpretation (Sussmann, 1957; Heisenberg, 1958), according to which the quantum state description is not supposed to express the properties of a physical system or ensemble of systems but our *knowledge* of these properties, and changes of the state (such as the supposed reduction of state during measurement) are identified with changes of knowledge, not so much because it is wrong as because it is irrelevant.[18] One can, of course, study a person's knowledge of physics rather than studying physics itself, but such a study is not germane to this paper. For example, someone's knowledge of a certain system can change discontinuously as a result of a blow to the head which causes amnesia, as well as through the receipt of new information from a

[14] The tensor product notation $A\otimes B$ means that A operates on the $\mid\text{I}\rangle$ vectors, and B operates on the $\mid\text{II}\rangle$ vectors.

[15] This illustrates the fact that the state of a two-component system cannot be uniquely determined by measurements on each component separately. Jauch (1968, Secs. 11-7 and 8) gives a good discussion of the mathematical aspects of this problem, and in particular of the tensor product formalism.

[16] As pointed out in DLP (1966), this is not necessarily the thermodynamic definition of metastability.

[17] On this point Rosenfeld seems to have modified his earlier views, as expressed in a discussion exchange with Viger (see Körner, 1957, pp. 183–6).

[18] Another area in which the subjective interpretation causes confusion is the relationship between entropy and information. It is true that the aquisition of information requires a corresponding increase of entropy (Brillouin, 1962); hence the association of information with negative entropy is useful. But the converse proposition that entropy is nothing but a measure of our lack of information is a fallacy. Entropy also has a thermodynamic meaning, $dS=dQ/T$, which is valid regardless of the existence or nonexistence of Maxwell's demon.

measurement, although the subjectivists tend to ignore the former while stressing the latter.

5. JOINT PROBABILITY DISTRIBUTIONS

In contrast to the previous sections, which were concerned entirely with interpretation of the existing formalism (summarized in Sec. 1.1), this section and the next consider possible extensions of the formalism. These will not be modifications of the established formalism, but additions to it which are compatible with both the established formalism and the Statistical Interpretation.

As discussed in Sec. 3 with regard to the uncertainty principle, quantum theory is not inconsistent with the supposition that a particle has at any instant both a definite position and a definite momentum, although there is a widespread folklore to the contrary. Of course we are not compelled either to accept or to reject this supposition, but it is of interest to explore it on a tentative basis.

Our first problem is to construct, for the noncommuting observables q and p, a joint probability distribution $P(q, p; \psi)$ for any state $|\psi\rangle$,[19] such that the single variable distributions constructed from it agree with the established formalism, i.e.,

$$\int P(q, p; \psi)\, dp \equiv P(q; \psi) = |\langle\psi | q\rangle|^2, \quad (5.1)$$

$$\int P(q, p; \psi)\, dq \equiv P(p; \psi) = |\langle\psi | p\rangle|^2. \quad (5.2)$$

Additional conditions will be considered later.

Let us first consider the method by which the single observable probability distribution function can be constructed. The characteristic function for the distribution of an arbitrary observable A is given by

$$M(\xi; \psi) = \langle \exp(i\xi A)\rangle = \int \exp(i\xi A) P(A; \psi)\, dA, \quad (5.3)$$

hence $P(A; \psi)$ is equal to the inverse Fourier transform of $M(\xi; \psi)$. By expanding the exponential

$$M(\xi; \psi) = \sum_{n=0}^{\infty} \frac{(i\xi)^n}{n!} \langle A^n\rangle, \quad (5.4)$$

we see that a knowledge of the characteristic function is equivalent to a knowledge of all the moments of the distribution. According to the formalism (see F4), these are given by[20]

$$\langle A^n\rangle = \langle\psi | \hat{A}^n | \psi\rangle. \quad (5.5)$$

By analogy we may introduce a characteristic function for the joint probability distriubtion

$$M(\theta, \lambda; \psi) = \sum_{n,m=0}^{\infty} \frac{(i\theta)^n (i\lambda)^m}{n!m!} \langle q^n p^m\rangle, \quad (5.6)$$

[19] We shall consider only pure states since these illustrate the essential problems. The generalization to mixed states is straightforward.

[20] In this section it is necessary to distinguish between a physical observable and the mathematical operator which represents it. The operators will be distinguished by a circumflex.

from which, by a Fourier inversion, we obtain,

$$P(q, p; \psi) = (2\pi)^{-2} \int\int M(\theta, \lambda; \psi) \exp[-i(\theta q + \lambda p)]\, d\theta d\lambda. \quad (5.7)$$

As long as the moments $\langle q^n\rangle$ and $\langle p^m\rangle$ are given by (5.5), then (5.7) will automatically satisfy (5.1) and (5.2).

The difficulty now evident is that that there is no unique way in which to take products of noncommuting operators. For example q^2p^2 may be represented by $\frac{1}{2}(\hat{q}^2\hat{p}^2 + \hat{p}^2\hat{q}^2)$, $\frac{1}{4}(\hat{q}\hat{p} + \hat{p}\hat{q})^2$, $\frac{1}{2}(\hat{q}\hat{p}\hat{q}\hat{p} + \hat{p}\hat{q}\hat{p}\hat{q})$, or any of several other forms. In these examples we have already used the additional plausible restrictions that the product be a self-adjoint operator, and that it be symmetric under interchange of $\hat{q}$ and $\hat{p}$. Shewell (1959) has considered several correspondence rules which have been proposed, and found none to be fully satisfactory. Some of the rules do not yield a unique operator, while others yield an operator for a power of the Hamiltonian, $[H(q, p)]^n$, which is not equal to $(\hat{H})^n$. Each such possible choice will yield a different joint probability distribution which satisfies (5.1) and (5.2).

If one chooses

$$\exp[i(\theta q + \lambda p)] \rightarrow \exp(i\theta\hat{q} + i\lambda\hat{p}), \quad (5.8a)$$

or equivalently

$$q^n p^m \rightarrow 2^{-n} \sum_{l=0}^{n} \binom{n}{l} \hat{q}^{n-l} \hat{p}^m \hat{q}^l, \quad (5.8b)$$

then one obtains (Moyal, 1949)

$$P(q, p; \psi) = (2\pi)^{-1} \int \langle\psi | q - \tfrac{1}{2}\tau\hbar\rangle \exp(-i\tau p) \langle q + \tfrac{1}{2}\tau\hbar | \psi\rangle\, d\tau, \quad (5.9)$$

which was first introduced by Wigner (1932). Because this expression may become negative it cannot be interpreted as a genuine probability distribution, and Wigner proposed it only as a calculational device. Another distribution which suffers from the same defect has been derived by Margenau and Hill (1961) using the correspondence

$$q^n p^m \rightarrow \tfrac{1}{2}(\hat{q}^n \hat{p}^m + \hat{p}^m \hat{q}^n). \quad (5.10)$$

Since every correspondence rule which associates an operator with a classical function of q and p,

$$g(q, p) \rightarrow \hat{G}(\hat{q}, \hat{p}), \quad (5.11)$$

provides a joint probability distribution, it is possible to formulate the most general form of $P(q, p; \psi)$ which will satisfy (5.1) and (5.2). We refer to the original papers for details (Cohen, 1966; Margenau and Cohen, 1967). These authors investigated the possibility of constructing a joint probability distribution such that

$$\langle\psi | \hat{G}(\hat{q}, \hat{p}) | \psi\rangle = \int\int g(q, p) P(q, p; \psi)\, dq dp, \quad (5.12)$$

and also such that for any function K

$$\langle\psi \mid K(\hat{G}(\hat{q}, \hat{p})) \mid \psi\rangle=\iint K(g(q, p)) P(q, p ; \psi)\, d q d p. \tag{5.13}$$

In other words, they asked whether there exists a joint probability distribution such that the averages of all observables can be calculated by a phase-space average as in classical statistical mechanics.[21]

Their answer was negative. Although it is always possible to satisfy (5.12), (5.13) could be simultaneously satisfied only if the correspondence

$$K(g(q, p)) \rightarrow K(\hat{G}(\hat{q}, \hat{p})) \tag{5.14}$$

can be satisfied for all functions K with the same correspondence rule as that leading to (5.11). This they prove to be impossible.[22] With the wisdom of hindsight, we should not be surprised that (5.12) and (5.13) cannot be satisfied, for if it were otherwise, then quantum mechanics could be expressed as a special case of classical mechanics.

To deal with simultaneous values of position and momentum variables in a logically consistent fashion we need only require that a joint probability distribution exists which satisfies (5.1) and (5.2), and

$$P(q, p ; \psi) \geq 0 \tag{5.15}$$

for all quantum states. These conditions are obviously satisfied by the function

$$P(q, p ; \psi)=|\langle\psi \mid q\rangle|^2|\langle\psi \mid p\rangle|^2, \tag{5.16}$$

but this form is probably not unique. Indeed, it cannot be applicable to scattering experiments (see Fig. 3), where, merely by geometry, there must be a close correlation between position and momentum at large distances from the scattering center. The investigation of possible joint probability distributions is clearly not complete.

Bopp (1956, 1957) has undertaken not only to represent quantum mechanics in terms of ensembles in phase space, but also to derive the statistical equations from simple principles—an *a priori* derivation—not merely a translation from the vector-operator formalism of quantum theory. Bopp's work differs from the above in that his joint probability distribution does not satisfy (5.1) and (5.2). Nevertheless, there is a definite relation between the averages calculated in Bopp's theory and those of quantum theory (denoted by subscripts B and Q). For example,

$$\langle q\rangle_B=\langle q\rangle_Q, \qquad \langle p\rangle_B=\langle p\rangle_Q,$$
$$\langle q^2\rangle_B=\langle q^2\rangle_Q+l^2/4,$$
$$\langle p^2\rangle_B=\langle p^2\rangle_Q+\hbar^2/l^2.$$

Here l is some constant with dimensions of length (interpreted by Bopp as the finest accuracy of a position measurement).

The problem of defining a satisfactory joint probability distribution for position and momentum may not be entirely mathematical, for we must also decide on "the empirical definition of the concept 'simultaneous values of position and momentum,' " according to Prugovecki (1967). To this end he studies various operational methods of measuring both these observables *simultaneously* on the same individual system, not merely on different individual systems representative of the same state preparation. He suggests that the irreducible errors in individual measurements ought to be taken into account in a generalized formalism (in addition to the statistical fluctuations in an ensemble of similar measurements, which are treated in the established formalism). It should be emphasized that simultaneous measurement of noncommuting observables, even with a finite precision, has no counterpart in the established formalism which treats only measurement of a single observable (or a commuting set of observables).[23] Now to experimentally determine a probability distribution, one must perform a number of measurements and construct a histogram of the results (see Fig. 2). Each point on the curve will have a vertical uncertainty due to the statistical uncertainty of using a finite sample, and a horizontal uncertainty due to the error involved in a single measurement. (In the case of simultaneous measurement of q and p, a single measurement would be a pair of numbers (q, p), and the error would be an area in the qp plane). If the latter cannot be reduced to zero, we cannot determine a probability distribution exactly, even if we eliminate statistical uncertainties by repeating the measurement an arbitrarily large number of times.

To deal with such a circumstance Prugovecki defines a *complex probability distribution*, $P_\psi^{q,p}(I_1 \times I_2)$, with the following properties: the real part of P represents the probability that q will be in the interval, I_1 and simultaneously p will be in the interval I_2, where

$$I_1=\{\lambda: q_1 \leq \lambda \leq q_2\},$$
$$I_2=\{\mu: p_1 \leq \mu \leq p_2\}, \tag{5.17}$$

and the imaginary part of P represents the uncertainty associated with the previous statement. Certain conditions must be satisfied in order for this interpretation to be sensible. Since q and p are certain to be somewhere in their spectra, we must have $\operatorname{Re} P \rightarrow 1$, and $\operatorname{Im} P \rightarrow 0$, as I_1 and I_2 are enlarged to include the entire spectra from $-\infty$ to $+\infty$. If I_1 and I_2 are smaller than the irreducible errors in a single measurement,

[21] Because (5.1) and (5.2) are satisfied, this will clearly be true for any function of q only, or of p only.

[22] It would seem inevitable that arbitrary functional relations involving q and p cannot be preserved by any quantal operator correspondence rule, since the equation $qp-pq=0$ for classical variables becomes $\hat{q}\hat{p}-\hat{p}\hat{q}=i\hbar$ in quantum theory. Clearly one cannot map zero onto $i\hbar$ in a consistent fashion.

[23] The naive interpretation (Matthews, 1968, Chap. 3) of the operator product AB as corresponding to a measurement of B followed by a measurement of A is refuted by the example of the Pauli spin operators for which $\sigma_x\sigma_y=i\sigma_z$. A measurement of σ_y followed by a measurement of σ_x is in no way equivalent to a measurement of σ_z.

then Im P should be very large because one cannot determine whether or not a measured value of (q, p) lies within such a small region of phase space, and therefore one cannot determine the relative frequency (probability) of such an event. Of course the complex probability distribution must reduce to the real probability distribution of the established formalism in the case of commuting operators.

The specific expression proposed by Prugovecki for the complex probability distribution is

$$P_\psi^{q,p}(I_1\times I_2)=\langle\psi \mid E_1(I_1)E_2(I_2)\mid\psi\rangle, \quad (5.18)$$

where E_1 and E_2 are projection operators which project onto the subspaces spanned by the eigenvectors of $\hat{q}$ corresponding to the spectral interval I_1, and of $\hat{p}$ corresponding to I_2, respectively. This form, although perhaps the most obvious choice, may not be satisfactory because he shows that for any nontrivial intervals I_1 and I_2 there exist states for which Re $P<0$. This would be tolerable if simultaneously we had a large value for Im P, so that we could say that the probability was not well defined in such a case. But in fact one obtains Im $P=0$ for that state which makes Re P most negative. Since a "well defined" but negative probability makes no sense, some modification of either (5.18) or the interpretation seems in order (such as, perhaps, a restriction on the states ψ which are to be considered physically realizable). This objection does not necessarily invalidate the concept of a complex probability distribution, but the usefulness of such a distribution is yet to be determined.

6. HIDDEN VARIABLES

Quantum theory predicts the statistical distribution of events (i.e., the results of similar measurements preceded by a certain state preparation). But if the prepared state does not correspond to an eigenvector of the particular observable being measured, then the outcome of any individual event is not determined by quantum theory. Thus one is led to the conjecture that the outcome of an individual event may be determined by some variables which are not described by quantum theory, and which are not controllable in the state preparation procedure. The statistical distributions of quantum theory would then be averages over these "hidden variables."

The entirely reasonable question, "Are there hidden-variable theories consistent with quantum theory, and if so, what are their characteristics?," has been unfortunately clouded by emotionalism. A discussion of the historical and psychological origins of this attitude would not be useful here. We shall only quote one example of an argument which is in no way extreme (Inglis, 1961, p. 4), "Quantum mechanics is so broadly successful and convincing that the quest [for hidden variables] does not seem hopeful." The vacuous character of this argument should be apparent, for the success of quantum theory within its domain of definition (i.e., the calculation of statistical distributions of events) has no bearing on the existence of a broader theory (i.e., one which could predict individual events).

The question of hidden variables has also been subject to a genuine confusion (i.e., not merely due to the conflicting personalities or metaphysical ideas of the principal characters in the debate), for Von Neumann proved a theorem from which he concluded that no hidden-variable theory could reproduce all of the statistical predictions of quantum theory. Although his mathematical theorem is correct, his physical conclusion is not, and the first example of a hidden-variable model was published by Bohm (1952). However, a clear analysis of the nature of Von Neumann's theorem and its relation to hidden variables was not achieved until much later (Bell, 1966), and the 14 year interval between these papers yielded many inconclusive discussions pro and con, as well as some "improved proofs" of the impossibility of hidden variables. We shall expound briefly the content of Von Neumann's theorem, and the reason why it does not rule out hidden variables in quantum theory. Other relevant theorems will also be discussed.

The Statistical Interpretation, which regards quantum states as being descriptive of ensembles of similarily prepared systems, is completely open with respect to hidden variables. It does not demand them, but it makes the search for them entirely reasonable [this was the attitude of Einstein (1949)]. On the other hand, the Copenhagen Interpretation, which regards a state vector as an exhaustive description of an individual system, is antipathetic to the idea of hidden variables, since a more complete description than that provided by a state vector would contradict that interpretation.

6.1 Von Neumann's Theorem

Von Neumann's mathematical theorem concerns the average of an observable, represented by a Hermitian operator, in a general ensemble subject only to a few conditions. The original derivation (Von Neumann, 1955, pp. 305–325) is lengthier than necessary, and we shall follow the more concise version given by Albertson (1961). Von Neumann makes the following essential assumptions which in his book are identified by the symbols in brackets:

(i) [I] If an observable is represented by the operator R, then a function f of that observable is represented by $f(R)$.

(ii) [II] The sum of several observables represented by $R, S, \cdots$ is represented by the operator $R+S+\cdots$, regardless of whether these operators are mutually commutative.

(iii) [p. 313, not identified] The correspondence between Hermitian operators and observables is one to one.

(iv) [A′] If the observable R is nonnegative, then $\langle R\rangle\geq 0$.

(v) [B′] For arbitrary observables R, S, $\cdots$ and arbitrary real numbers a, $b\cdots$, we must have

$$\langle aR+bS+\cdots\rangle=a\langle R\rangle+b\langle S\rangle+\cdots \tag{6.1}$$

for all possible ensembles (or states) in which the averages may be calculated.

An arbitrary Hermitian operator R may be written as

$$\begin{aligned}R&=\sum_{m,n}|n\rangle R_{nm}\langle m|\\&=\sum_n U^{(n)}R_{nn}\\&\quad+\sum_{m,n;m<n}\{V^{(nm)}\operatorname{Re}R_{mn}+W^{(nm)}\operatorname{Im}R_{mn}\},\end{aligned}\tag{6.2}$$

where $R_{nm}=R_{mn}{}^{*}=\langle n|R|m\rangle$ is a number, and

$$\begin{aligned}U^{(n)}&=|n\rangle\langle n|,\\V^{(nm)}&=|n\rangle\langle m|+|m\rangle\langle n|,\\W^{(nm)}&=-i(|n\rangle\langle m|-|m\rangle\langle n|)\end{aligned}\tag{6.3}$$

are Hermitian operators. According to assumption (iii) these operators are all to be regarded as observables, and hence (ii) and (v) may be applied to obtain

$$\begin{aligned}\langle R\rangle&=\sum_n\langle U^{(n)}\rangle R_{nn}\\&\quad+\sum_{m,n;m<n}\{\langle V^{(nm)}\rangle\operatorname{Re}R_{mn}+\langle W^{(nm)}\rangle\operatorname{Im}R_{mn}.\end{aligned}\tag{6.4}$$

This result may be rewritten as

$$\langle R\rangle=\sum_{m,n}\rho_{nm}R_{mn}=\operatorname{Tr}(\rho R),\tag{6.5}$$

if we define the matrix elements of the statistical operator ρ to be

$$\begin{aligned}\rho_{nn}&=\langle U^{(n)}\rangle,\\\rho_{mn}&=\tfrac{1}{2}\langle V^{(nm)}\rangle+\tfrac{1}{2}i\langle W^{(nm)}\rangle\qquad(m<n),\\\rho_{nm}&=\tfrac{1}{2}\langle V^{(nm)}\rangle-\tfrac{1}{2}i\langle W^{(nm)}\rangle\qquad(m<n).\end{aligned}\tag{6.6}$$

Since the averages of observables are real, the operator ρ is Hermitian by construction, and assumption (iv) implies that it is nonnegative definite ($\langle\phi|\rho|\phi\rangle\geq 0$ for all ϕ).

This completes the theorem of Von Neumann, which may be summarized by saying that the statistical operator representation of states (F4 of our Sec. 1.1) need not be introduced as a postulate, but may instead be derived from assumptions (i)$\cdots$(v). This theorem, which is clearly of interest outside of the question of hidden variables, is discussed in more detail in Chapter 5 of Jordan (1969).

Hidden variables are, by definition, hypothetical variables whose values must be specified in addition to the values of a complete commuting set of observables in order to uniquely determine the result of any measurement on a system. Quantum states, with their characteristic statistical distributions, would represent ensembles of systems with different values of the hidden variables, but if all hidden variables were fixed, the resultant subquantal ensemble would be dispersionless. That is,

$$\langle(R-\langle R\rangle)^2\rangle=0,\tag{6.7a}$$

or equivalently

$$\langle R^2\rangle=\langle R\rangle^2,\tag{6.7b}$$

for all observables R, since by hypothesis every quantity would have a unique value. Applying the result (6.5) to (6.7b) for the case of $R=|\phi\rangle\langle\phi|$, where $\langle\phi|\phi\rangle=1$, yields

$$\langle\phi|\rho|\phi\rangle=\langle\phi|\rho|\phi\rangle^2\tag{6.8}$$

for all normalized vectors $|\phi\rangle$. This implies $\langle\phi|\rho|\phi\rangle=1$ or 0 for all $|\phi\rangle$, and since this expression varies continuously with $|\phi\rangle$, the constant—1 or 0—must be the same for all $|\phi\rangle$. Hence we must have either

$$\rho=1\quad\text{or}\quad\rho=0\tag{6.9}$$

for dispersion-free ensembles.

The case $\rho=0$ is ruled out because it would imply $\langle R\rangle=0$ for all R in any subquantal ensemble, and so averages of these could never yield the correct averages in quantum ensembles. The case $\rho=1$ does not in fact yield a dispersionless ensemble if the vector space is greater than one dimensional. In this case

$$\langle 1\rangle=\operatorname{Tr}(1)=d,\tag{6.10}$$

d being the dimensionality of the space (usually ∞), and the left-hand side of (6.7a) becomes

$$\langle R^2\rangle-2\langle R\rangle^2+\langle R\rangle^2\langle 1\rangle\tag{6.11}$$

which is not zero. Thus Von Neumann concludes that, provided his assumptions (i)–(v) are accepted, there are no dispersion-free ensembles, and hence there can be no hidden variables. Since these assumptions are generally considered to be a part of quantum theory, he states "It is therefore not, as is often assumed, a question of a reinterpretation of quantum mechanics,—the present system of quantum mechanics would have to be objectively false $\cdots$," in order for even hypothetical hidden variables to be introduced in a logically consistent fashion.

The above conclusion is incorrect, but before we discuss the reasons why, let us dispose of two minor points of confusion. Albertson (1961, 1962) makes misleading and incorrect statements to the effect that Von Neumann's theorem does not *assume* the existence of noncommuting observables, and that the theorem is independent of, and additional to, the uncertainty principle. But the assumption (iii), which Von Neumann did not distinguish by any symbol, in effect assumes the existence of an infinite number of noncommuting observables, and this assumption plays a very central role in the proof.[24] If one assumed that all observables

[24] Oddly enough, Jammer (1966) repeats Albertson's incorrect statement on p. 369, while on p. 370 he mentions Von Neumann's assumption that every projection operator is an observable, without being aware of the contradiction!

were commutative, then a simultaneous eigenvector of all the observables would represent a dispersion-free ensemble.

The second minor point concerns the observation that (iii) is trivially false if there are superselection rules, which divide the full Hilbert space into coherent subspaces such that no observable may have matrix elements between vectors of different subspaces and no physical state vector may be a superposition of vectors from different subspaces (Wick *et al.*, 1952; Galindo *et al.*, 1962). But Von Neumann's theorem can still be proven in each coherent subspace individually, so the question of superselection rules is irrelevant.

6.2 Bell's Rebuttal

There is nothing wrong with the mathematics of Von Neumann's theorem. Nevertheless his conclusion that no hidden-variable model can be consistent with the statistical predictions of quantum theory is false—for such models exist (Bohm, 1952). The difficulty lies in the relation between the mathematics and the physics, as was clearly analyzed by Bell (1966).

The result of Von Neumann's theorem [Eq. (6.5)] seems to imply that any ensemble can be characterized by a statistical operator, whether this ensemble is a quantum state or a subquantal hidden-variable state. But since we know that a statistical operator characterizes just a general quantum state, we should immediately become suspicious of the assumptions which lead to this result. As Bell pointed out, it is (6.1) of assumption (v) which is at fault. At first sight the requirement that the average of the sum of two observables be equal to the sum of the averages of the observables separately may seem reasonable. But the nontrivial nature of this additivity property becomes apparent when we realize that it is not true for individual measurements of noncommuting observables. Consider for example the spin components of a particle. The measurement of σ_x can be made with a suitably oriented Stern–Gerlach magnet. The measurement of σ_y requires another orientation. There is no way of relating a measurement of $(\sigma_x+\sigma_y)$ to the results of the first two measurements. This requires a third and different orientation of the magnet. Moreover, the result of the measurement, an eigenvalue of $(\sigma_x+\sigma_y)$, will *not* be the sum of an eigenvalue of σ_x plus an eigenvalue of σ_y. That the ensemble average of many measurements of $(\sigma_x+\sigma_y)$, $\langle(\sigma_x+\sigma_y)\rangle$, should be equal to the sum of the averages of two other measurements involving different experimental arrangements, $\langle\sigma_x\rangle+\langle\sigma_y\rangle$, is a peculiar and nontrivial property of quantum states. But in a hypothetical dispersion-free state every observable would have a unique value equal to one of its eigenvalues, and since there is no linear relationship between the eigenvalues of noncommuting observables, in general, it is obvious that (6.1) could not possibly be satisfied for dispersion-free states. When this impossible condition is removed it is possible to construct a hidden-variable model which reproduces the correct statistical distributions for quantum states, as Bell showed by means of a simple example.

Bell also considers the relevance of some more recent mathematical work by Jauch and Piron (1968) and by Gleason (1957) to the problem of hidden variables, and we refer the reader to his paper for very clear discussions. Although the mathematical theorems of Von Neumann and the above authors do not in fact exclude hidden variables from quantum theory, they are not entirely devoid of value. With proper analysis, such as Bell provided, they help to point out some of the features which a successful hidden-variable model must possess.

The work of Gleason (1957) is particularly interesting, for he shows that the additivity assumption (6.1) for *commuting observables only* is incompatible with the requirement that every projection operator have a unique value (either 0 or 1) in a dispersion-free state. Bell shows that one may still introduce hidden variables by invoking the dependence of a measurement result on the "whole experimental arrangement" (Bohr, 1949, p. 222) in a very strong way. In his model the particular result obtained when measuring a certain observable may depend upon which other commuting observables are being measured simultaneously.

One undesirable feature of Gleason's work, in common with that of Von Neumann, is the assumption that every projection operator represents an observable. Since every Hermitian operator is a linear combination of commuting projection operators, this is essentially equivalent to the assumption that every Hermitian operator represents an observable (see F7 of Sec 1.1). We are not concerned with the comparatively trivial restrictions on observability imposed by superselection rules, but with the question of whether an operator such as $x^2p_x^3x^2$ really represents an observable quantity. One might conjecture, as this author has, that the restrictions Gleason's theorem imposes on hidden variables might be met in a less drastic fashion than that proposed by Bell, if only essential observables, rather than all projection operators, were required to have unique values (eigenvalues) in a hypothetical dispersion-free state. The problem of enumerating "essential observables" has not been seriously considered, but an important example due to Bell (1964), to be discussed, indicates that such a proposal would not be adequate.

It seems appropriate to reply here to Jauch and Piron (1968) and Misra (1967), who have commented somewhat unfavorably on the work of Bell (1966). Misra (p. 845) states, "The quest for hidden variables becomes a meaningful scientific pursuit only if states, even physically unrealizable states, are somehow restricted by physical condition, ···" with the implication that they have done this but that Bell's analysis is only "a drastic *ad hoc* modification of the notion of state" (Jauch and Piron, 1968). While agreeing with the first quoted statement above, I would contend that Misra's abstract algebraic approach, and the proposition-lattice theoretic approach of Jauch and Piron, fail to meet this requirement in that they impose only *abstract mathematical* conditions whose physical implications are obscured by their abstractness. The danger inherent in such an approach is underlined by the 34 year interval between Von Neumann's abstract attack on the hidden-variable problem and Bell's demonstration that one of his assumptions was physically unreasonable, and even impossible.

6.3 Bell's Theorem

Although hidden-variable models which reproduce all the statistical predictions of quantum theory are known to exist, they would be more interesting if they could be made to satisfy the very reasonable assumption (Einstein, 1949, p. 85) that the real factual situation of a system is independent of what is done with some other system which is spacially separated from, and not interacting with, the former. Bell (1964)[25] has considered this problem for the experiment which we discussed in connection with the EPR theorem, the measurement of arbitrary components of the spins of two, spin one-half particles, which were initially prepared in a singlet spin state, and which have separated.

The spin components $\boldsymbol{\sigma}_1\cdot\mathbf{a}$ and $\boldsymbol{\sigma}_2\cdot\mathbf{b}$ can both be measured by means of two Stern–Gerlach magnets oriented along the directions of the unit vectors **a** and **b**. The results of these two measurements will exhibit a statistical correlation, which for the singlet state is given by

$$\langle(\boldsymbol{\sigma}_1\cdot\mathbf{a})(\boldsymbol{\sigma}_2\cdot\mathbf{b})\rangle=-\mathbf{a}\cdot\mathbf{b}, \qquad (6.12)$$

according to quantum theory.

We now suppose that the results of individual measurements, which are not determined by the quantum state, are determined by some set of parameters denoted λ. The result A of measuring $\boldsymbol{\sigma}_1\cdot\mathbf{a}$ is determined by **a** and λ, and the result B of measuring $\boldsymbol{\sigma}_2\cdot\mathbf{b}$ is determined by **b** and λ; these results necessarily being eigenvalues of the operators, i.e.,

$$A(\mathbf{a},\lambda)=\pm1, \qquad B(\mathbf{b},\lambda)=\pm1. \qquad (6.13)$$

But in accordance with Einstein's assumption, the result A of measurement on particle 1 should not depend on the direction **b** of the magnet which acts on particle 2, and B should not depend on **a**. If $\rho(\lambda)$ is the probability distribution of the hidden variables λ, then the average of the product $(\boldsymbol{\sigma}_1\cdot\mathbf{a})(\boldsymbol{\sigma}_2\cdot\mathbf{b})$ will be given by

$$P(\mathbf{a},\mathbf{b})=\int A(\mathbf{a},\lambda)B(\mathbf{b},\lambda)\rho(\lambda)\,d\lambda. \qquad (6.14)$$

Bell then proves that (6.14) cannot be arbitrarily close to the result of quantum theory (6.12). To avoid the possibility that the differences might occur only at isolated points, he first averages (6.12) and (6.14) over small cones of angles about the mean directions of **a** and **b**. We may write

$$|\,\overline{\mathbf{a}\cdot\mathbf{b}}-\mathbf{a}\cdot\mathbf{b}\,|\leq\delta, \qquad (6.15)$$

where the bar denotes a smoothed function, with δ tending to zero with size of the range of angular averaging. Suppose that the difference between the smoothed functions is bounded,

$$|\,\bar{P}(\mathbf{a},\mathbf{b})+\overline{\mathbf{a}\cdot\mathbf{b}}\,|\leq\epsilon. \qquad (6.16)$$

From a straightforward argument based on (6.13)–(6.16), Bell deduces the inequality

$$4(\epsilon+\delta)\geq\sqrt{2}-1, \qquad (6.17)$$

from which it follows that as the range of angular smoothing, and hence δ, is made arbitrarily small, ϵ cannot be arbitrarily small. Hence (6.14) cannot be an arbitrarily accurate approximation to the quantum theory result (6.12).

Bell shows that it is easy to make a hidden-variable model to agree with (6.12) if the result $A(\mathbf{a},\lambda)$ of the measurement on particle 1 is allowed to depend also upon the direction **b** of the other magnet. However, if Einstein's hypothesis of independence of separated noninteracting systems is accepted, then the above theorem demonstrates that no such hidden-variable model can agree with all the predictions of quantum theory. This result can be interpreted as an illustration of Gleason's theorem as analysed by Bell (1966), according to which a hidden-variable model is possible only if the result of a measurement of $\boldsymbol{\sigma}_1\cdot\mathbf{a}$ depends upon which one of the observables $\boldsymbol{\sigma}_2\cdot\mathbf{b}$ (out of all the possible directions **b**) is being measured simultaneously. This example appears to rule out the possibility, conjectured in the previous section, that one might avoid the consequences of Gleason's theorem by narrowing one's attention from all Hermitian operators to only "essential observables," for there seems no doubt that $\boldsymbol{\sigma}_1\cdot\mathbf{a}$ and $\boldsymbol{\sigma}_2\cdot\mathbf{b}$ for all **a** and **b** are genuine observables.

Note added in proof: Recently Wigner (1970) has given a somewhat simpler version of Bell's argument, which yields some insight into its mathematical nature.

We shall not discuss specific hidden-variable theories in detail because none seems to be at all definitive, and any that reproduce *all* the results of quantum theory *exactly* must, as a consequence of Bell's theorem, be physically unreasonable (though they may be logically self-consistent). This remark would not apply to theories which depart from the formalism of quantum theory and only agree with it approximately in some limit. For descriptions of specific hidden-variable theories we refer the reader to a review by Freistadt (1957), and an article by Bohm (1962).

[25] In spite of the publication dates, it is clear from the contexts that this paper was written later than Bell (1966).

6.4 Suggested Experiments

As emphasized by Bohm, the postulates of quantum theory can be tested only if we consider what it would mean to contradict them. Therefore, at the very least, the study of hidden variables will have served a useful purpose if it leads to the suggestion of interesting experiments, regardless of whether or not the outcome of those experiments is favorable to the hidden-variable hypothesis.

The additivity of averages of noncommuting observables in quantum ensembles (6.1) is a very powerful assumption, as Von Neumann's theorem shows, since it and a few other postulates allow one to derive (6.5) from which all the statistical predictions of quantum theory flow. Von Neumann (1955, p. 309, Footnote 164) suggested an experiment based upon the fact that the Hamiltonian operator is the sum of two noncommuting terms, the kinetic and potential energies. For some suitable state, one should measure the average kinetic energy by measurements of the momentum, measure the average potential energy by measurements of position, and measure the average energy by spectroscopic methods (each measurement being preceded by the same state preparation). Then the additivity of the quantum averages (expectation values) could be directly tested. A similar experiment which would probably be easier to perform could be based on the spin components σ_x, σ_y, and $(\sigma_x+\sigma_y)=\sqrt{2}$ (component of $\boldsymbol{\sigma}$ at 45° to the x and y axes).

The experiment involving the correlation of spin components of two particles from an initial singlet state is of interest because Bell (1964) showed that, while it is easy to construct a hidden-variable theory to reproduce the results of quantum theory for measurements on a *single* spin, it is not possible to do so for two-spin correlations without violating Einstein's assumption. Moreover, from (6.17) we see that the hidden-variable prediction must differ from the result of quantum theory by at least $\epsilon > 0.10$, so a clear cut experimental test is possible.

A measurement of the correlation in the polarizations of photons emitted from the decay of singlet positronium, which is similar in principle to the experiment above, has been performed by Wu and Shaknov (1950), but only for two angles. As Bell (1964) has pointed out, it is easy to make a hidden-variable theory agree with quantum theory at certain fixed angles, but not at all angles.

The hidden-variable theory of Bohm and Bub (1966) has been tested by Papaliolios (1967). According to the theory, when a photon from an arbitrarily linearly polarized beam is incident on a Polaroid filter (whose polarization axis is not exactly parallel to the polarization vector of the beam), the photon is or is not transmitted depending upon the value of some hidden variable. In a normal beam, the hidden variables are uniformly distributed, yielding the usual transmission probability. In the experiment, the beam is first passed through two nearly crossed polaroid filters. According to the theory, the transmitted beam should then have almost unique values for the hidden variables. These are presumed to relax once more to a uniform distribution within some relaxation time τ, but if the beam is incident on a third rotatable polarizer (whose axis is at a variable angle θ to the axis of the second), before this relaxation takes place the intensity transmitted through it is predicted to deviate from the $\cos^2\theta$ law in a definite fashion. In the experiment, the third polarizer was placed as close to the second as possible, but no deviations from the $\cos^2\theta$ law were observed. Papaliolios concludes that the hidden variables, if they exist, must relax to a uniform distribution in a time $\tau < 2.4\times10^{-14}$ sec, which is two orders of magnitude smaller than the value suggested by Bohm and Bub.

7. CONCLUDING REMARKS

The central theme connecting all of the topics in this paper has been the superiority of a purely statistical interpretation of quantum theory, in which a state vector represents an *ensemble* of similarly prepared systems, as opposed to the stronger assumption that a state vector provides a complete description of an *individual* system. By the criterion of logical economy, the Statistical Interpretation is preferable because it makes fewer assumptions. The stronger assumption plays no role in any application of quantum theory, and so Occam's razor may be invoked to discard it.

A more serious argument against the stronger assumption above is that it leads to conceptual difficulties. One such difficulty is exhibited by the Theorem of Einstein, Podolsky, and Rosen (Sec. 2) which demonstrates that this assumption is incompatible with the physical independence of spacially separated objects which may have interacted in the past. Although from a purely logical point of view one could retain either of these assumptions, it seems most unreasonable to discard the independence of separated noninteracting objects.

It used to be argued that the peculiarities consequent upon the assumption that a system is completely characterized by a wave function were confined to the microscopic domain, and that these peculiarities were justified by "the unavoidable disturbance of the system by the measuring apparatus." The latter of these contentions is refuted by the example due to EPR, in which it is possible to measure some observable of an elementary particle, without the apparatus interacting with that particle in any way. It appears (see Witmer, 1967, p. 45) that Bohr recognized this implication of the EPR argument, for he subsequently cautioned against such phrases as "disturbing of phenomena by observation" (Bohr, 1949, p. 237), but this lesson has been forgotten by some modern authors (e.g., Messiah, 1964, p. 140; Matthews, 1968, p. 27).

The former contention is refuted by an analysis of a measurement of a microscopic object by a macroscopic apparatus (Sec. 4). It follows from very simple considerations that the final state of the system (object plus apparatus) *must*, in general, be a coherent superposition of macroscopically distinct "pointer positions." If the state vector is assumed to *completely* describe the individual system, i.e., if one assumes that the system simply does not have a value for any observable except those for which the state is an eigenvector (or perhaps those whose dispersion is very small), then one will be forced to the absurd conclusion that the "pointer" of the apparatus (a *macroscopic* object) has no definite position. The supposed *reduction of the state vector*, and all the difficulties and complications associated with it, are only artifacts of the vain attempt to retain the above assumption. But to what purpose? Under the more modest assumption that a state vector represents an *ensemble* of similarily prepared systems, the measurement process poses no particular problem.

The criticisms above also apply to certain hidden-variable theories in which a ψ field is associated with each individual system (Bohm, 1952; Bohm and Bub, 1966). Like Von Neumann's theory, these theories must also invoke a reduction of the ψ field upon measurement, which in the example of the correlated but spacially separated systems considered by EPR, would necessitate a signal passing instantaneously across the macroscopic distances between the systems. Since, as we have seen, such a drastic assumption is *not necessary* for a satisfactory interpretation of quantum theory, we see no reason for considering it further.

The *Uncertainty Principle* (Sec. 3) finds its natural interpretation as a lower bound on the statistical dispersion among similarily prepared system (this interpretation being *deduced*, not introduced *ad hoc*), and is not in any real sense related to the possible disturbance of a system by a measurement. The distinction between *measurement* and *state preparation* is essential for clarity. It is possible to extend the formalism of quantum theory by the introduction of *joint probability distributions* for position and momentum (Sec. 5). This demonstrates that there is no conflict with quantum theory in thinking of a particle as having definite (but, in general, unknown) values of both position and momentum, contrary to an earlier interpretation of the uncertainty principle.

The Statistical Interpretation does not prejudice the possibility of introducing *hidden variables* which would determine (in principle) the outcome of each individual measurement (Sec. 6). Although such models exist, Bell (1964) showed that they cannot, in general, satisfy the requirement that measurements on spacially separated noninteracting objects should be independent. It is ironic that this requirement which Einstein, Podolsky, and Rosen first used to refute the belief that a wave function could completely describe an individual system (a belief which stood opposed to hidden variables), should also prove disasterous to at least the simplest idea of hidden variables.

Many of the ideas expounded in this paper were expressed, in essence, by Einstein (1949) in his *Reply to Criticisms*. In support of his thesis that a wave function must be considered to represent an ensemble of systems, and cannot reasonably be assumed to provide a complete description of an individual system, he considers the decay of a radioactive atom, with an automatic recording detector to register the decay time.[26] This example is a prototype of the one used in the Theory of Measurement, and it embodies all the essential physical content of the more general argument. In this *Reply* he also restated the conclusion of the EPR argument in the form which we have called the Theorem of EPR, and gave a brief discussion of the nontrivial aspect of the classical limit of the quantal state description.

A serious reading of Einstein's *Reply* should clear up any misconceptions to the effect that he rejected quantum theory or misunderstood its foundations. In fact, he understood the essentially *statistical* nature of quantum theory as well as any of his contemporaries, and better than many. His only objection was against the assumption that a wave function or state vector could exhaustively describe an *individual* system, which we have seen to be an unwarranted and troublesome assumption. This fact, and the fact that Einstein advocated a fully viable interpretation of quantum theory (essentially the Statistical Interpretation of this paper although he expressed himself more briefly), do not seem to have been appreciated by his critics.[27]

The foundations of quantum theory are subject to continuous discussion, and two almost coincident papers in the *American Journal of Physics*, taking nearly opposite points of view, deserve comment. Hartle (1968) has made a novel attempt to derive the statistical assertions of quantum theory from a quantal description of individual systems. The conceptual basis of this attempt is questionable, for, as Hartle admits, a quantum state is *not* an objective property of an individual system. That is to say, in the words of Blokhintsev (1968, p. 50),

"If [the wave function were a characteristic of a single particle] it would be of interest to perform such a measurement on a single particle (say an electron) which would allow us to determine its own individual wave function. No such measurement is possible."

But whereas Blokhintsev (and the present author) regards the quantum state as describing an *ensemble* of similarily prepared systems, Hartle considers it to describe the information possessed by some observer. The irrelevance of that interpretation has already been commented upon (Sec. 4.4).

[26] Messiah's (1964, p. 158) reply to this argument is essentially that there are experimental arrangements from which it is impossible to determine a definite decay time. (His wording to the effect that they are *incompatible* with the *existence* of a time of decay is an unjustified overstatement). But this is irrelevant. The point is that it is *sometimes* (i.e., with suitable apparatus) possible to measure the decay time of a single atom, but this decay time is *never* predicted by quantum theory.

[27] See the articles by Pauli, Born, Heitler, Bohr, and Margenau in the volume edited by Schilpp (1949). However, Margenau's recent papers are more compatible with Einstein's views.

On the other hand Park (1968) has carefully studied the relations between *ensembles* and *individual* systems in statistical theories. He shows how one may proceed logically from the description of ensembles to individual systems in classical statistical mechanics, and how the attempt to proceed similarily in quantum statistical mechanics to an identification of a *pure state* operator with an individual system fails. The present paper is consistent with his conclusion; moreover we have shown that such an identification would lead to serious difficulties in principle.

Although this article has been concerned primarily with a consistent interpretation of quantum theory as it is presently formulated, one is naturally led to ask for hints of future developments. Except for eliminating some false starts, such as theories of wave function reduction, such hints are hard to find. Recognition that quantum states should refer to ensembles of similarily prepared systems would seem to open the door for hidden variables to control individual events. On the other hand, Bell's theorem seems to present a severe obstacle for any hidden-variable theory which reproduces exactly *all* of the predictions of quantum theory, but this need not be so for a theory which departs from the formalism of quantum theory and recovers it only in some limiting case. Presumably the next step must be a bold departure from the familiar formalism, as Einstein's theory of gravitation departed from that of Newton.

REFERENCES

Albertson, J., 1961, Am. J. Phys. **29**, 478.
——, 1962, Brit. J. Phil. Sci. **13**, 229.
Allcock, G. R., 1969, Ann. Phys. (N.Y.) **53**, 253.
Araki, H., and M. M. Yanase, 1960, Phys. Rev. **120**, 622.
Bell, J. S., 1964, Physics (N.Y.) **1**, 195.
——., 1966, Rev. Mod. Phys. **38**, 447
Blokhintsev, D. I., 1968, *The Philosophy of Quantum Mechanics*, (Reidal, Dordrecht, Holland). Original Russian edition, 1965.
Bohm, D., 1951, *Quantum Theory* (Prentice-Hall, Englewood Cliffs, N.J.).
——, 1952, Phys. Rev. **85**, 166, 180.
——, 1953, Phys. Rev. **89**, 458.
——, 1962, in *Quantum Theory*, edited by D. R. Bates (Academic, New York) Vol. 3.
——, and Y. Aharonov, 1957, Phys. Rev. **108**, 1070.
——, and J. Bub, 1966, Rev. Mod. Phys. **38**, 453.
Bohr, N., 1935, Phys. Rev. **48**, 696.
——, 1949, in *Albert Einstein: Philosopher–Scientist*, edited by P. A. Schilpp (Library of Living Philosphers, Evanston, Ill.), reprinted by Harper and Row, New York, p. 199.
Bopp, F., 1956, Ann. Inst. Henri Poincaré, **15**, 81.
——, 1957, in *Observation and Interpretation*, edited by S. Korner (Butterworths, London), p. 189.
Born, M., 1926, Z. Physik **37**, 863.
Brillouin, L., 1962, *Science and Information Theory* (Academic, New York).
Breitenberger, E., 1965, Nuovo Cimento **38**, 356.
Bunge, M., 1956, Am. J. Phys. **24**, 272.
Carruthers, P., and M. M. Nieto, 1968, Rev. Mod. Phys. **40**, 411.
Cohen, L., 1966, J. Math. Phys. **7**, 781.
Daneri, A., A. Loinger, and G. M. Prosperi, 1962, Nucl. Phys. **33**, 297.
——, A. Loinger, and G. M. Prosperi, 1966, Nuovo Cimento **44B**, 119.
Day, T. B., 1961, Phys. Rev. **121**, 1204.
Dirac, P. A. M., 1958, *The Principles of Quantum Mechanics* (Clarendon, Oxford), 4th edition.
Duane, W., 1923, Proc. Natl. Acad. Sci. U.S. **9**, 158.
Einstein, A., B. Podolsky, and N. Rosen, 1935, Phys. Rev. **47**, 777.
——, 1949, in *Albert Einstein: Philosopher–Scientist*, edited by P. A. Schilpp (Library of the Living Philosophers, Evanston, Ill.), reprinted by Harper and Row, New York, p. 665.
Fano, U., 1957, Rev. Mod. Phys. **29**, 74.
Feyerabend, P. K., 1957, in *Observation and Interpretation*, edited by S. Körner (Butterworths, London), p. 121.
Freistadt, H., 1957, Nuovo Cimento Suppl. **5**, 1–70.
Furry, W. H. 1936, Phys. Rev. **49**, 393.
Galindo, A., A. Morales, and R. Nunez-Lagos, 1962, J. Math. Phys. **3**, 324.
Gleason, A. M., 1957, J. Math & Mech. **6**, 885.
Hartle, J. B., 1968, Am. J. Phys. **36**, 704.
Heisenberg, W., 1927, Z. Physik **43**, 172.
——, 1930, *The Physical Principles of Quantum Theory* (University of Chicago Press, Chicago, Ill.), reprinted by Dover, New York, 1950.
——, 1955, in *Neils Bohr and the Development of Physics*, edited by W. Pauli (Pergamon, London), p. 12.
——, 1958, *Physics and Philosophy* (Harper & Brothers, New York), pp. 54–55.
Heitler, W., 1949, in *Albert Einstein: Philosopher–Scientist*, edited by P. A. Schilpp (Library of the Living Philosophers, Evanston, Ill.), reprinted by Harper and Row, New York, p. 179.
Inglis, D. R., 1961, Rev. Mod. Phys. **33**, 1.
Jammer, M., 1966, *The Conceptual Development of Quantum Mechanics* (McGraw-Hill, New York).
Jauch, J. M., 1964, Helv. Phys. Acta **37**, 293.
——, 1968, *Foundations of Quantum Mechanics* (Addison–Wesley, Reading, Mass.).
——, and C. Piron, 1968, Rev. Mod. Phys. **40**, 228.
Jordan, T. F., 1969, *Linear Operators for Quantum Mechanics* (Wiley, New York).
Judge, D., and J. T. Lewis, 1963, Phys. Letters **5**, 190.
Komar, A., 1962, Phys. Rev. **126**, 365.
Korner, S., Ed., 1957, *Observation and Interpretation* (Butterworths, London).
Lamb, W. E., 1969, Phys. Today **22**, No. 4, 23.
Landé, A., 1955, *Foundations of Quantum Theory* (Yale U. P., New Haven, Conn.).
——, 1965, *New Foundations of Quantum Theory* (Cambridge U. P., Cambridge, England).
Margenau, H., 1958, Phil. Sci. **25**, 23.
——, 1963, Ann. Phys. (N.Y.) **23**, 469.
——, and L. Cohen, 1967, in *Quantum Theory and Reality*, edited by M. Bunge (Springer, New York), p. 71.
——, and R. N. Hill, 1961, Progr. Theoret. Phys. (Kyoto) **26**, 722.
Matthews, P. T., 1968, *Introduction to Quantum Mechanics* (McGraw-Hill, London).
Messiah, A., 1964, *Quantum Mechanics* (North-Holland, Amsterdam).
Misra, B., 1967, Nuovo Cimento **47A**, 841.
Mott, N. F., 1964, Contemporary Phys. **5**, 401.
Moyal, J. E., 1949, Proc. Cambridge Phil. Soc. **45**, 99.
Papaliolios, C., 1967, Phys. Rev. Letters **18**, 622.
Park, J. L., 1968, Am. J. Phys. **36**, 211.
Peierls, R. E., 1967, Sci. Am. **216**, No. 1, 137.
Popper, K. R., 1959, *The Logic of Scientific Discovery* (Basic Books, New York).
——, 1967, in *Quantum Theory and Reality*, edited by M. Bunge (Springer, New York), p. 7.
Prugovečki, E., 1967, Can J. Phys. **45**, 2173.
Robertson, H. P., 1929, Phys. Rev. **34**, 163.
Rosenfeld, L., 1965, Progr. Theoret. Phys. (Kyoto) Suppl., extra number, p. 222 (Yukawa commemorative issue).
——, 1968, Nucl. Phys. **A108**, 241.
Schilpp, P. A., 1949, Ed. *Albert Einstein: Philosopher–Scientist* (Library of the Living Philosophers, Evanston, Ill.), reprinted by Harper and Row, New York, 1959.
Schrödinger, E., 1930, Preuss. Akad. Wiss. Berlin, Ber. **19**, 296.
——, 1935, Naturwiss. **23**, 807.
Sharp, D. H., 1961, Phil. Sci. **28**, 225.
Shewell, J. R. 1959, Am. J. Phys. **27**, 16.
Shimony, A., 1966, Phys. Today **19**, No. 9, 85.
Susskind, L., and J. Glogower, 1964, Physics (N.Y.) **1**, 49.
Sussman, G., 1957 in *Observation and Interpretation*, edited by S. Korner (Butterworths, London), p. 131.

Tisza, L., 1963, Rev. Mod. Phys. **35**, 151.
Van der Waerden, B. L., 1967, *Sources of Quantum Mechanics* (North-Holland, Amsterdam).
Von Neumann, J., 1955, *Mathematical Foundations of Quantum Mechanics* (Princeton U. P., Princeton, N.J.). This is a translation of *Mathematische Grundlagen der Quantenmechanik* (Springer, Berlin, 1932).
Wakita, H., 1960, Progr. Theoret. Phys. (Kyoto) **23**, 32.
——, 1962, Progr. Theoret. Phys. (Kyoto) **27**, 139, 1156.
Whitten, R. C., 1966, Am. J. Phys. **34**, 1203.
Wick, G. C., A. S. Wightman, and E. P. Wigner, 1952, Phys. Rev. **88**, 101.
Wigner, E. P., 1932, Phys. Rev. **40**, 749.
——, 1962, in *The Scientist Speculates*, edited by I. J. Good (Basic Books, New York), p. 284.
——, 1963, Am. J. Phys. **31**, 6.
——, 1970, Am. J. Phys. **38**, 1005.
Witmer, E. E., 1967, Am. J. Phys. **35**, 40.
Wu, C. S., and I. Shaknov, 1950, Phys. Rev. **77**, 136.
Yanase, M. M., 1961, Phys. Rev. **123**, 666.

AJP Volume 40

The Copenhagen Interpretation*

HENRY PIERCE STAPP
Lawrence Berkeley Laboratory
University of California
Berkeley, California 94720
(Received 26 March 1971; revised 29 February 1972)

An attempt is made to give a coherent account of the logical essence of the Copenhagen interpretation of quantum theory. The central point is that quantum theory is fundamentally pragmatic, but nonetheless complete. The principal difficulty in understanding quantum theory lies in the fact that its completeness is incompatible with external existence of the space–time continuum of classical physics.

I. INTRODUCTION

Scientists of the late twenties, led by Bohr and Heisenberg, proposed a conception of nature radically different from that of their predecessors. The new conception, which grew out of efforts to comprehend the apparently irrational behavior of nature in the realm of quantum effects, was not simply a new catalog of the elementary space–time realities and their modes of operation. It was essentially a rejection of the presumption that nature could be understood in terms of elementary space–time realities. According to the new view, the complete description of nature at the atomic level was given by probability functions that referred not to underlying microscopic space–time realities but rather to the macroscopic objects of sense experience. The theoretical structure did not extend down and anchor itself on fundamental microscopic space–time realities. Instead it turned back and anchored itself in the concrete sense realities that form the basis of social life.

This radical concept, called the Copenhagen interpretation, was bitterly challenged at first but became during the '30's the orthodox interpretation of quantum theory, nominally accepted by almost all textbooks and practical workers in the field.

Recently, perhaps partly in response to the severe technical difficulties now besetting quantum theory at the fundamental level, there has been mounting criticism of the Copenhagen interpretation. The charges range from the claim that it is a great illogical muddle to the claim that it is in any case unnecessary, and hence, in view of its radical nature, to be rejected. Reference 1 contains some stoutly worded attacks on the Copenhagen interpretation. Reference 2 is a more moderately worded review article that firmly rejects the Copenhagen interpretation. Reference 3 is a list of recent articles in the physical literature that espouse a variety of views on the question.

The striking thing about these articles is the diversity they reveal in prevailing conceptions of the Copenhagen interpretation itself. For example, the picture of the Copenhagen interpretation painted in Ref. 1 is quite different from the pictures painted in Refs. 2 and 3 by practicing physicists. And these latter pictures themselves are far from uniform.

The cause of these divergences is not hard to find. Textbook accounts of the Copenhagen interpretation generally gloss over the subtle points. For clarification the readers are directed to the writings of Bohr[4] and Heisenberg.[5] Yet clarification is difficult to find there. The writings of Bohr are extraordinarily elusive. They rarely seem to say what you want to know. They weave a web of words around the Copenhagen interpretation but do not say exactly what it is. Heisenberg's writings are more direct. But his way of speaking suggests a subjective interpretation that appears quite contrary to the apparent intentions of Bohr. The situation is perhaps well summarized by von Weizsäcker, who, after expressing the opinion that the Copenhagen interpretation is correct and indispensable, says he must "add that the interpretation, in my view, has never been fully clarified. It needs an interpretation, and that will be its only defense."[6]

von Weizsäcker is surely correct. The writings of Bohr and Heisenberg have, as a matter of historical fact, not produced a clear and unambiguous picture of the basic logical structure of their position. They have left impressions that vary significantly from reader to reader. For this reason a clarification of the Copenhagen interpretation is certainly needed. My aim here is to provide one. More precisely, my aim is to give a clear account of the logical essence of the Copenhagen interpretation. This logical essence should be distinguished from the inhomogeneous body of opinions and views that now constitute the Copenhagen interpretation itself. The logical essence constitutes, I believe, a completely rational and coherent position.

The plan of the work is as follows. First, quantum theory is described from the point of view of actual practice. Then, to provide contrast, several non-Copenhagen interpretations are considered. Next, to provide background, some philosophical ideas of William James are introduced. The pragmatic character of the Copenhagen interpretation is then discussed, and the incompatibility of the completeness of quantum theory with the external existence of the space–time continuum of classical physics is noted. Finally, the question of the completeness of quantum theory is examined.

II. A PRACTICAL ACCOUNT OF QUANTUM THEORY

Quantum theory is a procedure by which scientists predict probabilities that measurements of specified kinds will yield results of specified kinds in situations of specified kinds. It is applied in circumstances that are described by saying that a certain physical system is first prepared in a specified manner and is later examined in a specified manner. And this examination, called a measurement, is moreover such that it can yield, or not yield, various possible specified results.

The procedure is this: The specifications A on the manner of preparation of the physical system are first transcribed into a wave function $\Psi_A(x)$. The variables x are a set of variables that are characteristic of the physical system being prepared. They are called the degrees of freedom of the prepared system. The description of the specifications A is couched in a language that is meaningful to an engineer or laboratory technician. The way in which these operational specifications A are translated into a corresponding wave function $\Psi_A(x)$ is discussed later.

The specifications B on the subsequent measurement and its possible result are similarly couched in a language that allows a suitably trained technician to set up a measurement of the specified kind and to determine whether the result that occurs is a result of the specified kind. These specifications B on the measurement and its result are transcribed into a wave function $\Psi_B(y)$, where y is a set of variables that are called the degrees of freedom of the measured system.

Next a transformation function $U(x;\, y)$ is constructed in accordance with certain theoretical rules. This function depends on the type of system that was prepared and on the type of system that was measured, but not on the particular wave functions $\Psi_A(x)$ and $\Psi_B(y)$. The "transition amplitude"

$$\langle A \mid B \rangle \equiv \int \Psi_A(x) U(x;\, y) \Psi_B{}^*(y)\, dx\, dy$$

is computed. The predicted probability that a measurement performed in the manner specified by B will yield a result specified by B, if the preparation is performed in the manner specified by A, is given by

$$P(A,\, B) = |\langle A \mid B \rangle|^2.$$

The experimental physicist will, I hope, recognize in this account a description of how he uses quantum theory. First he transforms his information about the preparation of the system into an initial wave function. Then he applies to it some linear transformation, calculated perhaps from the Schrödinger equation, or perhaps from the S matrix, which converts the initial wave function into a final wave function. This final wave function, which is built on the degrees of freedom of the measured system, is then folded into the wave function corresponding to a possible result. This gives the transition amplitude, which is

Henry Pierce Stapp

multiplied by its complex conjugate to give the predicted transition probability.

In a more sophisticated calculation one might use density matrices $\rho_A(x';x'')$ and $\rho_B(y';y'')$ instead of $\Psi_A(x)$ and $\Psi_B(y)$ to represent the prepared system and the possible result. This would allow for preparations and measurements that correspond to statistical mixtures. But this generalization could be obtained also by simply performing classical averages over various $\Psi_A(x)$ and $\Psi_B(y)$.

The above account describes how quantum theory is used in practice. The essential points are that attention is focused on some system that is first prepared in a specified manner and later examined in a specified manner. Quantum theory is a procedure for calculating the predicted probability that the specified type of examination will yield some specified result. This predicted probability is the predicted limit of the relative frequency of occurrence of the specified result, as the number of systems prepared and examined in accordance with the specifications goes to infinity.

The wave functions used in these calculations are functions of a set of variables characteristic of the prepared and measured systems. These systems are often microscopic and not directly observable. No wave functions of the preparing and measuring devices enter into the calculation. These devices are described operationally. They are described in terms of things that can be recognized and/or acted upon by technicians. These descriptions refer to the macroscopic properties of the preparing and measuring devices.

The crucial question is how does one determine the transformations $A \rightarrow \Psi_A$ and $B \rightarrow \Psi_B$. These transformations transcribe procedural descriptions of the manner in which technicians prepare macroscopic objects, and recognize macroscopic responses, into mathematical functions built on the degrees of freedom of the (microscopic) prepared and measured systems. The problem of constructing this mapping is the famous "problem of measurements" in quantum theory.

The problem of measurements was studied by von Neumann.[7] He begins with the idea that one should describe the combined system composed of the original systems plus the original measuring devices in terms of a quantum mechanical wave function, and use quantum theory itself to calculate the needed mappings. This program has never been carried out in any practical case. One difficulty is that actual macroscopic devices are so complicated that qualitative calculations lie beyond present capabilities. The second problem is that such calculations would, in any case, provide only connections between the wave functions ϕ of the preparing and measuring devices and the wave functions Ψ of the original system. There would remain the problem of finding the mappings $A \rightarrow \phi_A$ and $B \rightarrow \phi_B$.

von Neumann's approach is not the one that is adopted in actual practice; no one has yet made a qualitatively accurate theoretical description of a measuring device. Thus what experimentalists do, in practice, is to *calibrate* their devices.

Notice, in this connection, that if one takes N_A different choices of A and N_B different choices of B, then one has only $N_A + N_B$ unknown functions Ψ_A and Ψ_B, but $N_A \times N_B$ experimentally determinable quantities $|\langle A \mid B \rangle|^2$. Using this leverage, together with plausible assumptions about smoothness, it is possible to build up a catalog of correspondences between what experimental physicists do and see, and the wave functions of the prepared and measured systems. It is this body of accumulated empirical knowledge that bridges the gap between the operational specifications A and B and their mathematical images Ψ_A and Ψ_B.

The above description of how quantum theory is used in practice will be used in the account of the Copenhagen interpretation. Before describing that interpretation itself I shall, to provide contrast, describe several other approaches.

III. SEVERAL OTHER APPROACHES

A. The Absolute-Ψ Approach

von Neumann's lucid analysis of the process of measurement is the origin of much of the current worry about the interpretation of quantum theory. The basic worrisome point can be illustrated by a simple example.

Suppose a particle has just passed through one of two slits. And suppose a 100% efficient counter is placed behind each slit, so that by seeing which counter fires a human observer can determine through which slit the particle passed.

Suppose the particle is represented initially by a wave function that assigns equal probabilities to the parts associated with the two slits. And consider a quantum theoretical analysis of the process of measurement in which both the particle and the two counters are represented by wave functions.

It follows directly and immediately from the superposition principle (i.e., linearity) that the wave function of the complete system after the measurement necessarily will consist of a superposition of two terms. The first term will represent the situation in which (1) the particle has passed through the first counter, (2) the first counter has fired, and (3) the second counter has not fired. The second term will represent the situation in which (1) the particle has passed through the second counter, (2) the second counter has fired, and (3) the first counter has not fired. These two terms evolve from the two terms in the wave function of the initial particle. The presence of both terms is a direct and unavoidable consequence of the superposition principle, which ensures that the sum of any two solutions of the equation of motion is another solution.

Notice now that the counters are macroscopic objects and that the wave function necessarily contains a sum of two terms one of which corresponds to the first counter's having fired but not the second, and the other of which corresponds to the second counter's having fired but not the first. Thus the wave function necessarily corresponds to a sum of two logically incompatible *macroscopic* possibilities.

To dramatize this situation, suppose the human observer now looks at the counters and runs upstairs or downstairs depending on which counter he sees firing. Then the wave function of the entire system of particle plus counters plus human observer will consist, eventually, of a sum of two terms. One term will represent the human observer running upstairs, and the other term will represent this same human observer running downstairs. Both terms must necessarily be present in the wave function, simply by virtue of the superposition principle.

This fact that the wave function necessarily develops into a sum of parts that correspond to incompatible *macroscopic* possibilities must be squared with the empirical facts. The human observer does not run both upstairs and downstairs. He does one or the other, not both. Therefore the wave function must collapse to a form that is consistent with what actually does happen. But such a collapse is definitely incompatible with the superposition principle.

This violation of the superposition principle bothers some thinkers. Wigner[8] calls the existence of the two modes of change of the wave function—i.e., the smooth causal evolution and the fitful statistical jumps associated with measurements—a strange dualism, and says that the probabilistic behavior is almost diametrically opposite to what one would expect from ordinary experience. He and Ludwig[9] speculate that quantum theory may have to be modified by the addition of a nonlinear effect in the macroscopic realm in order to arrive at a consistent theory of measurements. Wigner[10] even speculates that the nonlinearity may be associated with the action of mind on matter.

An even more radical proposal was made by Everett[11] and supported by Wheeler[12] and Bryce DeWitt.[13] According to this proposal the human observer actually runs both upstairs and downstairs at the same time. When the human observer sees the counter fire he breaks into two separate editions of himself, one of which runs upstairs while the other runs down. However, the parts of the wave function corresponding to these two different possibilities move into different regions of the multiparticle configuration space and consequently do not interfere. Therefore the two editions will never be aware of each other's existence. Thus appearances are saved without violating the superposition principle.

This proposal is, I think, unreasonable. A wave function times its complex conjugate, has the mathematical properties of a probability function. Probability functions for composite systems are naturally defined on the product of the spaces of the individual component systems; it is this property that allows different statistical weights to be assigned to the various logically alternative possibilities. A decomposition of a wave function into parts corresponding to different logical alternatives is thus completely natural. In the example described—with the initial specification as described there—there is a finite probability that the observer will be running upstairs, and a finite probability that he will be running downstairs. Thus the wave function necessarily must have both parts. If it collapsed to one part or the other it would no longer correctly describe the probabilities corresponding to the original specifications.

Of course, if the original specifications are replaced by new ones that include now the specification that the observer is running upstairs, not downstairs, then the original wave function will naturally be replaced by a new one, just as it would be in classical statistical theory.

In short, the mathematical properties of the wave functions are completely in accord with the idea that they describe the evolution of the *probabilities* of the actual things, not the actual things themselves. The idea that they describe also the evolution of the actual things themselves leads to metaphysical monstrosities. These might perhaps be accepted if they were the necessary consequences of irrefutable logic. But this is hardly the case here. The basis of Everett's whole proposal is the premise that the superposition principle cannot suddenly fail. This premise is sound. But the natural and reasonable conclusion to draw from it is that the wave functions describe the evolution of the *probabilities* of the actual things, not the evolution of the actual things themselves. For the mathematical form and properties of the wave function, including its lawful development in accordance with the superposition principle, are completely in accord with the presumption that it is a probability function. The addition of the metaphysical assumption that the wave function represents the evolution of not only the probabilities of the actual things, but of also the actual things themselves, is unreasonable because its only virtue is to save the superposition principle, which, however, is not in jeopardy unless one introduces this metaphysical assumption.

Everett's proposal, and also those of Wigner and Ludwig, are the outgrowth of a certain tendency to ascribe to the wave function a quality of absoluteness that goes beyond what is normally and naturally attached to a probability function. This tendency can perhaps be traced to what Rosenfeld[14] calls "a radical difference in conception (going back to von Neumann)...," this radical difference being with the ideas of Bohr. von Neumann's application of quantum theory to the process of measurement itself, coupled with his parallel treatments of the two very different modes of development of the wave function—i.e., the smooth dynamical evolution, and the abrupt changes associated with measurement—tend to conjure up the image of some absolute wave function developing in time under the influence of two different dynamical mechanisms. The living, breathing scientist who changes the wave function he uses as he receives more information is replaced by a new dynamical mechanism. The resulting picture is strange indeed.

In the Copenhagen interpretation the notion of an absolute wave function representing the world itself is unequivocally rejected. Wave functions, like the corresponding probability functions in classical physics, are associated with the studies by scientists of finite systems. The devices that prepare and later examine such systems are regarded as parts of the ordinary classical physical world. Their space–time dispositions are interpreted by the scientist as information about the prepared and examined systems. Only these latter systems are represented by wave functions. The probabilities involved are the probabilities of specified responses of the measuring devices under specified conditions.

New information available to the scientist can be used in two different ways. It can be considered to be information about the response of a measuring device to the system being examined. In this case the probability of this response is the object of interest. On the other hand, the new information can also be regarded as part of the specification of a new preparation. The wave function that represents this new specification will naturally be different from the wave function that represented the original specifications. One would not expect the superposition principle to be maintained in the change of the wave function associated with a change of specifications.

This pragmatic description is to be contrasted with descriptions that attempt to peer "behind the scenes" and tell us what is "really happening." Such superimposed images can be termed metaphysical appendages insofar as they have no testable consequences. The pragmatic interpretation ignores all such metaphysical appendages.

The sharp distinction drawn in this section

between probabilities and the actual things to which they refer should not be construed as an acceptance of the real-particle interpretation which is described next.

B. The Real-Particle Interpretation

The real-particle interpretation affirms that there are real particles, by which is meant tiny localized objects, or disturbances, or singularities, or other things that stay together like particles should, and do not spread out like waves. According to this interpretation, the probability functions of quantum theory describe, typically, the probability that a real particle is in such-and-such a region. This real-particle interpretation is defended by Popper in Ref. 1, and by Ballentine in Ref. 2.

Confidence in the existence of real particles was restored by Bohm's[15] illustration of how nonrelativistic Schrödinger theory can be made compatible with the existence of point particles. The price paid for this achievement is this: All the particles in the (model) universe are instantly and forcefully linked together. What happens to any particle in the universe instantly and violently affects every other particle.

In such a situation it is not clear that we should continue to use the term "particle." For the entire collection of "particles" in Bohm's universe acts as a single complex entity. Our usual idea of a particle is an abstraction from experience about macroscopic objects, and it normally carries, as part of the idea of localization, the idea that the localized entity is an independent entity, in the sense that it depends on other things in the universe only through various "dynamical" effects. These dynamical effects are characterized by a certain respect for space–time separations. In particular, they are "causal." If the connections between particles radically transcend our idea of causal dynamical relationships, then the appropriateness of the word "particle" can be questioned.

Recently, Bell[16] has shown that the statistical predictions of quantum theory are definitely incompatible with the existence of an underlying reality whose spatially separated parts are independent realities linked only by causal dynamical relationships. The spatially separated parts of any underlying reality must be linked in ways that completely transcend the realm of causal dynamical connections. The spatially separated parts of any such underlying reality are not independent realities, in the ordinary sense.

Bell's theorem does not absolutely rule out the real-particle interpretation, if one is willing to admit these hyperdynamical connections. But they fortify the opinion that a dynamical theory based on such a real entity would have no testable dynamical consequences. For the strong dependence of individual effects here on Earth upon the fine details of what is happening all over the universe apparently rules out any ordinary kind of test of such a theory.

IV. THE PRAGMATIC CONCEPTION OF TRUTH

To prepare the mind for the Copenhagen interpretation it is useful to recall some ideas of William James.[17] James argued at length for a certain conception of what it means for an idea to be true. This conception was, in brief, that an idea is true if it works.

James's proposal was at first scorned and ridiculed by most philosophers, as might be expected. For most people can plainly see a big difference between whether an idea is true and whether it works. Yet James stoutly defended his idea, claiming that he was misunderstood by his critics.

It is worthwhile to try to see things from James's point of view.

James accepts, as a matter of course, that the truth of an idea means its agreement with reality. The questions are: What is the "reality" with which a true idea agrees? And what is the relationship "agreement with reality" by virtue of which that idea becomes true?

All human ideas lie, by definition, in the realm of experience. Reality, on the other hand, is usually considered to have parts lying outside this realm. The question thus arises: How can an idea lying inside the realm of experience agree with something that lies outside? How does one conceive of a relationship between an idea, on the one hand, and something of such a fundamentally different sort? What is the structural form of that connection between an idea and a transexperiential reality that goes by the name of "agreement"? How can such a relationship be comprehended by thoughts forever confined to the realm of experience?

The contention that underlies James's whole position is, I believe, that a relationship between an idea and something else can be comprehended only if that something else is also an idea. Ideas are eternally confined to the realm of ideas. They can "know" or "agree" only with other ideas. There is no way for a finite mind to comprehend or explain an agreement between an idea and something that lies outside the realm of experience.

So if we want to know what it means for an idea to agree with a reality we must first accept that this reality lies in the realm of experience.

This viewpoint is not in accord with the usual idea of truth. Certain of our ideas are ideas about what lies outside the realm of experience. For example, I may have the idea that the world is made up of tiny objects called particles. According to the usual notion of truth this idea is true or false according to whether or not the world really is made up of such particles. The truth of the idea depends on whether it agrees with something that lies outside the realm of experience.

Now the notion of "agreement" seems to suggest some sort of similarity or congruence of the things that agree. But things that are similar or congruent are generally things of the same kind. Two triangles can be similar or congruent because they are the same kind of thing: The relationships that inhere in one can be mapped in a direct and simple way into the relationships that inhere in the other.

But ideas and external realities are presumably very different kinds of things. Our ideas are intimately associated with certain complex, macroscopic, biological entities—our brains—and the structural forms that can inhere in our ideas would naturally be expected to depend on the structural forms of our brains. External realities, on the other hand, could be structurally very different from human ideas. Hence there is no *a priori* reason to expect that the relationships that constitute or characterize the essence of external reality can be mapped in any simple or direct fashion into the world of human ideas. Yet if no such mapping exists then the whole idea of "agreement" between ideas and external realities becomes obscure.

The only evidence we have on the question of whether human ideas can be brought into exact correspondence with the essences of the external realities is the success of our ideas in bringing order to our physical experience. Yet the success of ideas in this sphere does not ensure the exact correspondence of our ideas to external reality.

On the other hand, the question of whether ideas "agree" with external essences is of no practical importance. What is important is precisely the success of the ideas—if ideas are successful in bringing order to our experience then they are useful even if they do not "agree," in some absolute sense, with the external essences. Moreover, if they are successful in bringing order into our experience then they do "agree" at least with the aspects of our experience that they successfully order. Furthermore, it is only this agreement with aspects of our experience that can ever really be comprehended by man. That which is not an idea is intrinsically incomprehensible, and so are its relationships to other things. This leads to the pragmatic viewpoint that ideas must be judged by their success and utility in the world of ideas and experience, rather than on the basis of some intrinsically incomprehensible "agreement" with nonideas.

The significance of this viewpoint for science is its negation of the idea that the aim of science is to construct a mental or mathematical image of the world itself. According to the pragmatic view, the proper goal of science is to augment and order our experience. A scientific theory should be judged on how well it serves to extend the range of our experience and reduce it to order. It need not provide a mental or mathematical image of the world itself, for the structural form of the world itself may be such that it cannot be placed in simple correspondence with the types of structures that our mental processes can form.

James was accused of subjectivism—of denying the existence of objective reality. In defending himself against this charge, which he termed slanderous, he introduced an interesting ontology consisting of three things: (1) private concepts, (2) sense objects, (3) hypersensible realities. The private concepts are subjective experiences. The sense objects are public sense realities, i.e., sense realities that are independent of the individual. The hypersensible realities are realities that exist independently of all human thinkers.[18]

Of hypersensible realities James can talk only obliquely, since he recognizes both that our knowledge of such things is forever uncertain and that we can moreover never even think of such things without replacing them by mental substitutes that lack the defining characteristics of that which they replace, namely the property of existing independently of all human thinkers.

James's sense objects are curious things. They are sense realities and hence belong to the realm of experience. Yet they are public: They are independent of the individual. They are, in short, objective experiences. The usual idea about experiences is that they are personal or subjective, not public or objective.

This idea of experienced sense objects as public or objective realities runs through James's writings. The experience "tiger" can appear in the mental histories of many different individuals. "That desk" is something that I can grasp and shake, and you also can grasp and shake. About this desk James says "But you and I are commutable here; we can exchange places; and as you go bail for my desk, so I can go bail for yours. This notion of a reality independent of either of us, taken from ordinary experience, lies at the base of the pragmatic definition of truth."[19]

These words should, I think, be linked with Bohr's words about classical concepts as the basis of communication between scientists. In both cases the focus is on the concretely experienced sense realities—such as the shaking of the desk—as the foundation of social reality. From this point of view the objective world is not built basically out of such airy abstractions as electrons and protons and "space." It is founded on the concrete sense realities of social experience, such as a block of concrete held in the hand, a sword forged by a blacksmith, a Geiger counter prepared according to specifications by laboratory technicians and placed in a specified position by experimental physicists.

This brief excursion into philosophy provides background for the Copenhagen interpretation, which is fundamentally a shift to a philosophic perspective resembling that of William James.

V. THE PRAGMATIC CHARACTER OF THE COPENHAGEN INTERPRETATION

The logical essence of the Copenhagen interpretation is summed up in the following two assertions:

(1) The quantum theoretical formalism is to be interpreted *pragmatically*.

(2) Quantum theory provides for a *complete* scientific account of atomic phenomena.

Point (1) asserts that quantum theory is *fundamentally* the procedure described in the practical account of quantum theory given in Sec. II. The central problem for the Copenhagen interpretation is to reconcile this assertion with the claim that it is complete, i.e., to reconcile assertions (1) and (2). This problem is discussed in Sec. VII.

The aim of the present section is to document point (1) by the words of Bohr. This fundamental point needs to be definitely settled, for critics often confuse the Copenhagen interpretation, which is basically pragmatic, with the diametrically opposed absolute-Ψ interpretation described in Sec. III. In what follows, particular attention will be paid to the possible conflict of the pragmatic viewpoint with (i) the element of realism in Bohr's attitude toward the macroscopic world and (ii) any commitment to a fundamental stochastic or statistical element in nature itself.

The quotations from Bohr that follow are taken from his three major works[4]: (I.) *Atomic Theory and the Description of Nature*; (II.) *Atomic Physics and Human Knowledge*; and (III.) *Essays 1958–1962 on Atomic Physics and Human Knowledge.*

The pragmatic orientation of the Copenhagen interpretation is fixed in the opening words of Bohr's first book: "The task of science is both to extend the range of our experience and reduce it to order ..." (I.1). "In physics ... our problem consists in the co-ordination of our experience of the external world ..." (I.1). "In our description of nature the purpose is not to disclose the real essence of phenomena but only to track down as far as possible relations between the multifold aspects of our experience" (I.18).

This commitment to a pragmatic view of science runs through all of Bohr's works. He later links it to the crucial problem of communication: "As the goal of science is to augment and order our experience every analysis of the conditions of human knowledge must rest on considerations of the character and scope of our means of communication" (II.88). "In this connection it is imperative to realize that in every account of physical experience one must describe both experimental conditions and observations by the same means of communication as one used in classical physics" (II.88). "The decisive point is to recognize that the description of the experimental arrangement and the recordings of observations must be given in plain language, suitably refined by the usual terminology. This is a simple logical demand, since by the word 'experiment' we can only mean a procedure regarding which we are able to communicate to others what we have done and what we have learnt" (III.3). "... we must recognize above all that, even when phenomena transcend the scope of classical physical theories, the account of the experimental arrangement and the recording of observations must be given in plain language, suitably supplemented by technical physical terminology. This is a clear logical demand, since the very word 'experiment' refers to a situation where we can tell others what we have done and what we have learned" (II.72).

Bohr's commitment to a pragmatic interpretation of the quantum-mechanical formalism is unambiguous: "...the appropriate physical interpretation of the symbolic quantum-mechanical formalism amounts only to predictions, of determinate or statistical character, pertaining to individual phenomena appearing under conditions defined by classical physical concepts" (II.64). "... the formalism does not allow pictorial representation on accustomed lines, but aims directly at establishing relations between observations obtained under well-defined conditions" (II.71). "The sole aim of (the quantum mechanical formalism) is the comprehension of observations obtained under experimental conditions described by simple physical concepts" (II.90). "Strictly speaking, the mathematical formalism of quantum mechanics and electrodynamics merely offers rules of calculation for the deduction of expectations about observations obtained under well-defined experimental conditions specified by classical physical concepts" (III.60).

Throughout Bohr's writings there is a tacit acceptance of the idea that the external world exists, and that our physical experiences are caused, in part, by the course of external events. This is quite in accord with pragmatism: James admits the existence of hypersensible realities. But there is no commitment by Bohr to the idea that the macroscopic world really is what we naively imagine it to be. The focus is on the *descriptions* of our physical experiences and the demand that they secure unambiguous communication and objectivity. Referring to the experimental arrangements and observations he says: "The description of atomic phenomena has in these respects a perfectly objective character, in the sense that no explicit reference is made to any individual observer and that therefore, with proper regard to relativistic exigencies, no ambiguity is involved in the communication of information. As regards all such points, the observation problem of quantum physics in no way differs from the classical physical approach" (III.3). Bohr's closest approach to a commitment to the idea that the macroscopic world actually is what it appears to be is, I think, the statement: "The renunciation of pictorial representation involves only the state of atomic objects, while the foundation of the description of the experimental conditions is fully retained" (II.90). The commitment here is, I believe, to the appropriateness, in quantum theory, of a classical *description* of the experimental conditions, rather than to the *fundamental* accuracy of classical ideas at the macroscopic level. This position is in complete accord with pragmatism.

In regard to the irreducible statistical element in quantum theory, Bohr was at first ambivalent. An initial acceptance of the notion of a fundamental element of randomness or indeterminism on the part of nature is suggested by the statement: "...we have been forced...to reckon with a free choice on the part of nature between various possibilities to which only probability interpretations can be applied" (I.4). However, he soon qualifies this idea (I.19) and later on says that at a Solvay conference

> an interesting discussion arose about how to speak of the appearance of phenomena for which only statistical predictions can be made. The question was whether, as to the occurrence of individual effects, we should adopt the terminology proposed by Dirac, that we were concerned with a choice on the part of "nature," or as suggested by Heisen-

> berg, we should say that we have to do with a choice on the part of the "observer" constructing the measuring instruments and reading their recording. Any such terminology would, however, appear dubious since, on the one hand, it is hardly reasonable to endow nature with volition in the ordinary sense, while on the other hand it is certainly not possible for the observer to influence the events which may appear under the conditions he has arranged. To my mind there is no other alternative than to admit in this field of experience, we are dealing with individual phenomena and that our possibilities of handling the measuring instruments allow us only to make a choice between the different complementary types of phenomena that we want to study (II.51).

Later he says "The circumstance that, in general, one and the same experimental arrangement may yield different recordings is sometimes picturesquely described as a 'choice of nature' between such possibilities. Needless to say, such a phrase implies no allusion to a personification of nature, but simply points to the impossibility of ascertaining on accustomed lines directives for the course of a closed indivisible phenomenon. Here, logical approach cannot go beyond the deduction of the relative probabilities for the appearance of the individual phenomena under given conditions" (II.73). "Corresponding to the fact that different individual quantum processes may take place in a given experimental arrangement these relations (between observations obtained under well-defined conditions) are of an inherently statistical character" (II.71). "The very fact that repetition of the same experiment, defined on the lines described, in general yields different recordings pertaining to the object, immediately implies that a comprehensive account of experience in this field must be expressed by statistical laws" (III.4). "The fact that in one and the same well-defined experimental arrangement we generally obtain recordings of different individual processes makes indispensible the recourse to a statistical account of quantum phenomena" (III.25). These statements indicate a turning away by Bohr from picturesque notions of a inherent random element in nature itself, and the adoption of an essentially pragmatic attitude toward the statistical character of the quantum-mechanical predictions.

It is worth noting that Bohr's notion of complementarity is altogether pragmatic: Ideas should be judged by their utility; physical ideas should be judged by their success in ordering physical experiences, not by the accuracy with which they can be believed to mirror the essence of external reality. The use of complementary ideas in complementary situations is a natural concomitant of pragmatic thinking.

VI. SPACE–TIME AND THE COMPLETENESS OF QUANTUM THEORY

In spite of doubts cast on our intuitive notions of space and time by the theory of relativity, the idea lingers on that persisting physical objects occupy space–time regions that can be divided into ever finer parts. A basic premise of classical physics is that this classical concept of the space–time continuum is the appropriate underlying concept for fundamental physical theory.

It is important to recognize that quantum theory has nothing in it that can be regarded as a description of qualities or properties of nature that are located at the point or infinitesimal regions of the space–time continuum. On one hand, the descriptions of the experimental arrangements and observations are basically operational descriptions of what technicians can see and do. They are not, strictly speaking, descriptions of the external things in themselves. Moreover, they are not descriptions of *microscopic* qualities or properties. On the other hand, the wave functions are merely abstract symbolic devices. They do not describe qualities or properties of nature that are located at points or infinitesimal regions of the space–time continuum. The abrupt change of a wave function in one region of space–time when a measurement is performed far away at the same time makes any such interpretation unreasonable. The wave functions of quantum theory are to be interpreted as symbolic devices that scientists use to make predictions about what they will observe under specified conditions. As Bohr says it: "In the treatment of atomic problems, actual calculations are most conveniently carried out with the help of a Schrödinger state function, from which the statistical laws governing observations attainable under specified conditions can be deduced by definite mathematical operations. It must be recognized, however, that we are here dealing with a purely symbolic procedure the unambiguous physical interpretation of which in the last resort requires a reference to the complete experimental arrangement" (III.5). "In fact, wave mechanics, just as the matrix theory, represents on this view a symbolic transcription of the problem of motion of classical mechanics adapted to the requirements of quantum theory and only to be interpreted by an explicit use of the quantum postulate" (I.75).

The fact that quantum theory contains nothing that is interpreted as a description of qualities located at points of an externally existing space–time continuum can be construed as evidence of its incompleteness. However, all we really know about the space–time continuum is that it is a concept that has been useful for organizing sense experience. Man's effort to comprehend the world in terms of the idea of an external reality inhering in a space–time continuum reached its culmination in classical field theory. That theory, though satisfactory in the domain of macroscopic phenomena, failed to provide a satisfactory account of the microscopic sources of the field. The bulk of Einstein's scientific life was spent in a frustrated effort to make these ideas work at the microscopic level.[20] The rejection of classical theory in favor of quantum theory represents, in essence, the rejection of the idea that external reality resides in, or inheres in, a space–time continuum. It signalizes the recognition that "space," like color, lies in the mind of the beholder.

If the classical concept of the space–time continuum were accepted, then quantum theory could not be considered complete, i.e., if it were accepted that the persisting objects of nature literally reside in a space–time continuum, with their various parts definitely located in specific regions, then a complete scientific account of atomic phenomena would, by virtue of the natural and normal meanings of these words, in this framework, be required to describe whatever it was that is located at the points or infinitesimal regions of that continuum. Quantum theory does not do this, and hence a claim of completeness would be an abuse of language.

In a pragmatic framework the claim of completeness has a different natural meaning. The natural meaning of the claim that quantum theory provides for a complete scientific account of atomic phenomena is that no theoretical construction can yield experimentally verifiable predictions about atomic phenomena that cannot be extracted from a quantum theoretical description. This is the practical or pragmatic meaning of scientific completeness in this context.

VII. WHOLENESS AND COMPLETENESS

The second essential ingredient of the Copenhagen interpretation is the claim that quantum theory provides for the complete scientific account of atomic phenomena. During the more than thirty years spanned by his three books[4] Bohr polished and refined his views on this point. His final, and I think best, summary is as follows:

> The element of wholeness, symbolized by the quantum of action and completely foreign to classical physical principles, has . . . the consequence that in the study of quantum processes any experimental inquiry implies an interaction between atomic object and the measuring tools which, although essential for the characterization of the phenomena, evades a separate account if the experiment is to serve its purpose of yielding unambiguous answers to our questions. It is indeed the recognition of this situation which makes recourse to a statistical mode of description imperative as regards to the expectations of the occurrence of individual quantum effects in one and the same experimental arrangement. (III.60).

This statement is augmented and clarified by an earlier statement:

> The essentially new feature in the analysis of quantum phenomena is . . . the introduction of a fundamental distinction between the measuring apparatus and the objects under investigation. This is a direct consequence of the necessity of accounting for the functions of the measuring instruments in purely classical terms, excluding in principle any regard to the quantum of action. On their side, the quantal features of the phenomena are revealed in the informa-

> tion about the atomic objects derived from the observations. While within the scope of classical physics the interaction between the object and apparatus can be neglected or, if necessary, compensated for, in quantum physics this interaction thus forms an inseparable part of the phenomena. Accordingly, the unambiguous account of proper quantum phenomena must, in principle, include a description of all relevant features of the experimental arrangement (III.3).

The basic point here is that well-defined objective specifications on the entire phenomenon are not restrictive enough to determine uniquely the course of the individual processes, yet no further breakdown is possible because of the inherent wholeness of the process symbolized by the quantum of action.

This way of tracing the need for a statistical account of atomic phenomena back to the element of wholeness symbolized by the quantum of action appears to take one outside the pragmatic framework since it refers to the measuring device, the atomic object, and their interaction. Also, it is not immediately clear how one is to reconcile the separate identification of these three things with the "impossibility of any sharp separation between the behaviour of the atomic objects and the interaction with the measuring instruments which serve to define the conditions under which the phenomena appear" (II.39).

In this connection it is important to recognize that the "atomic object" and "measuring instruments" are, within the framework of quantum thinking, *idealizations* used by scientists to bring order into man's experience in the realm of atomic phenomena. This point is developed by the author in Ref. 21. Bohr's words emphasize that these separate idealizations are inseparably linked by quantum thinking in a way that is completely foreign to classical thinking. The idealization "the measuring instrument" is a conceptual entity used in the description of the experimental specifications; the idealization "the atomic object" is a conceptual entity that is represented by the wave function. These are inseparably linked in quantum theory by the fact that the specifications described in terms of the measuring instrument are mapped onto wave functions associated with the atomic object: The atomic object represented by the wave function has no meaning in quantum theory except via its link to experience formulated in terms of specifications that refer to the measuring instruments.

Bohr evidently believed that there is in atomic processes an element of *wholeness*—associated with the quantum of action, and completely foreign to classical physical principles—that curtails the utility of the classical idealizations of the measuring instruments and atomic objects as separate, interacting entities, and that the resulting inseparability of the atomic object from the whole phenomenon renders statistical description unavoidable.

This way of reconciling the pragmatic character of quantum theory with the claim of completeness seems rational and coherent. It is, of course, based on quantum thinking itself and is therefore essentially a self-consistency consideration. The validity of quantum-mechanical thinking as a whole must, of course, be judged on the basis of its success, which includes its coherence and self-consistency.

The question of the completeness of quantum theory was debated by Bohr[22] and Einstein.[23] Einstein's counter arguments come down on the following points: (1) It is not proven that the usual concept of reality is unworkable; (2) quantum theory does not make "intelligible" what is sensorily given; and (3) if there is a more complete thinkable description of nature, then the formulation of the universal laws should involve their use.

Bell's theorem[16] deals a shattering blow to Einstein's position. For it proves that the ordinary concept of reality is incompatible with the statistical predictions of quantum theory. These predictions Einstein was apparently willing to accept. Einstein's whole position rests squarely on the presumption that sense experience can be understood in terms of an idea of some external reality whose spatially separated parts are independent realities, in the sense that they depend on each other only via connections that respect space–time separation in the usual way: Instantaneous connections are excluded. But the existence of such a reality lying behind the world of observed phenomena is precisely what Bell's theorem proves to be impossible.

Einstein's second point, about whether quantum theory makes intelligible what is sensorily given, is taken up in the next section.

Einstein's third point raises two crucial questions. The first is whether a complete description of nature is thinkable. Can human ideas, which are probably limited by the structural form of human brains, and which are presumably geared to the problem of human survival, fully know or comprehend the ultimate essences? And even if they can, what is the role in nature of universal laws? Is all nature ruled by some closed set of mathematical formulas? This might be one possibility. Another, quite compatible with present knowledge, is that certain aspects of nature adhere to closed mathematical forms but that the fullness of nature transcends any such form.

VIII. QUANTUM THEORY AND OBJECTIVE REALITY

The Copenhagen interpretation is often criticized on the grounds that it is subjective, i.e., that it deals with the observer's knowledge of things, rather than those things themselves. This charge arises mainly from Heisenberg's frequent use of the words "knowledge" and "observer." Since quantum theory is fundamentally a procedure by which scientists make predictions, it is completely appropriate that it refer to the knowledge of the observer. For human observers play a vital role in setting up experiments and in noting their results.

Heisenberg's wording, interpreted in a superficial way, can be, and has been, the source of considerable confusion. It is therefore perhaps better to speak directly in terms of the concrete social realities, such as dispositions of instruments, etc., in terms of which the preparations, measurements, and results are described. This type of terminology was favored by Bohr, who used the phrase "classical concepts" to signify descriptions in terms of concrete social actualities.

On the other hand, Bohr's terminology, though blatantly objective, raises the question of how quantum theory can be consistently constructed on a foundation that includes concepts that are fundamentally incompatible with the quantum concepts.

Perhaps the most satisfactory term is "specifications." Specifications are what architects and builders, and mechanics and machinists, use to communicate to one another conditions on the concrete social realities or actualities that bind their lives together. It is hard to think of a theoretical concept that could have a more objective meaning. Specifications are described in technical jargon that is an extension of everyday language. This language may incorporate concepts from classical physics. But this fact in no way implies that these concepts are valid beyond the realm in which they are used by the technicians.

In order to objectify as far as possible our descriptions of the specifications on preparations and measurements we can express them in terms of the "objective" quantities of classical physics. The meaning of these "objective" quantities for us is tied to the fact that we conceive of them as the qualities of an external world that exists independently of our perceptions of it. The formulation of the specifications in terms of these classical quantities allows the human observer to be eliminated, superficially at least, from the quantum theoretical description of nature: The observer need not be explicitly introduced into the description of quantum theory because the connection between his knowledge and these classical quantities is then shifted to other domains of science, such as classical physics, biology, psychology, etc.

But this elimination of the observer is simply a semantic sleight of hand. Since the conceptual structure of classical physics is recognized as fundamentally an invention of the mind that is useful for organizing and codifying experience, the knowledge of the observer emerges, in the end, as the fundamental reality upon which the whole structure rests. The terms "knowledge of the observer," or "classical description," or "specifications" are just different ways of summing up in a single term this entire arrangement of ideas, which follows from the recognition of the limited domain of validity of classical concepts.

Bohr cites certain ideas from biology and psychology as other examples of concepts that work well in certain limited domains. And he notes that there have been repeated attempts to unify all human knowledge of the basis of one or another of these conceptual frameworks.[24] Such attempts are the natural outgrowth of the absolutist viewpoint, which holds that the ideas of man can grasp or know the absolute essences. The pragmatist, regarding human concepts as simply tools for the comprehension of experience, and averring that human ideas, being prisoners in the realm of human experience, can "know" nothing but other human ideas, would not be optimistic about the prospects of complete success in such ventures. For him progress in human understanding would more likely consist of the growth of a web of interwoven complementary understandings of various aspects of the fullness of nature.

Such a view, though withholding the promise for eventual complete illumination regarding the ultimate essence of nature, does offer the prospect that human inquiry can continue indefinitely to yield important new truths. And these can be final in the sense that they grasp or illuminate some aspect of nature as it is revealed to human experience. And the hope can persist that man will perceive ever more clearly, through his growing patchwork of complementary views, the general form of a pervading presence. But this pervading presence cannot be expected or required to be a resident of the three-dimensional space of naive intuition, or to be described fundamentally in terms of quantities associated with points of a four-dimensional space–time continuum.

APPENDIX A: PHILOSOPHICAL ADDENDA

Several questions of a philosophic nature have been raised by a critic. This appendix contains my replies.

Question 1: How does one reconcile the commitment of James and Bohr to the public character of sense objects with the radical empiricist doctrine that ideas can agree only with other ideas? Russell's *Analysis of Matter* indicates the difficulty in performing this reconciliation.

Reply: Russell's arguments do not confute the ideas of James and Bohr as I have described them. Both of the latter authors would, I think, readily admit that human experiences are probably not the whole of reality but are probably merely a part of the whole that is related to the rest via some sort of causal-type connection. The critical question, however, is not whether there is in fact a world "out there," but rather what the connection is between the world "out there" and our ideas about it.

Russell argues, essentially, that we can make plausible inferences, based on the structure of our experiences, and build up a reasonable idea of the outside world. James would insist that this whole structure is nothing but a structure of abstract ideas built upon our common experiences, and that the transexperiential world that may somehow "cause" our common experiences never enters into this structure at all.

James evidently believes that his idea of a table is similar to yours and mine. In general, different people's ideas about sense objects are not identical, but they are similar enough to form the basis of effective social communication. There exists, in this sense, a realm of public or shared experiences that form the basis of interpersonal communication. This realm constitutes the primary data of science. The aim of science is to construct a framework of ideas that will link these common, or public, or shared, experiences together in ways that reflect various aspects of the empirical connections that exist between them. Thus the whole structure of science is, quite obviously, a structure that is wholly confined to the world of ideas.

Russell would presumably grant this. But he would argue that we can, nonetheless, make plausible inferences about the world based on the structure of experience. Yet his commitment to rationality requires him, I think, to admit that our ideas *might* not be able to fully comprehend the realities that are the causes of our experience. And if the evidence of science indicates that this possibility is the one realized by nature, then I think his rational approach, based on plausible inferences drawn from available evidence, would require him to admit that this possibility has a good "probability" of being correct.

Although the arguments of Russell do not confute the position of James, as I have described it, there is definitely a basic difference in orientation. Russell embarks on a quest for *certainty* about the external world, but settles for an account to which he assigns high "probability." James views the quest for certainty about the external world as totally misdirected. Certainty in such matters is clearly unattainable. The truly rational course of action is to admit at the outset that our aim is to construct a framework of ideas that is useful for the organization of our experience—and for the conduct of our lives—and to put aside the whole vague question about the connection of ideas to nonideas, and the equally vague question about the "probability" that a certain scheme of ideas gives us a true or valid picture of the world itself.

In any case, the claim that we *can* make valid inferences about the world itself acquires credibility only to the extent that a truly adequate picture of the world itself can be constructed. No such picture exists at present. And the difficulties in constructing a scientific view of the world itself are precisely those admitted by Russell himself, namely the incorporation of quantum phenomena and infinitesimal space–time intervals. It is precisely these difficulties that force us to fall back to the position of James.

In short, the position of Bohr and James, as I have described it, is not a denial of the causal theory of perception. It is simply a recommendation that we view science not as a quest for a metaphysical understanding of that which lies outside the world of ideas, but rather as an invention of the human mind that man constructs for the purpose of incorporating into the world of human ideas abstract structural forms that capture certain aspects of the empirical structure of man's experience. In this undertaking an important class of data are those experiences that are common to different human observers, such as our common or shared experience of the table about which we all sit. The level of experience at which these common experiences are most similar is the level at which a round table is experienced as a round table, not as an oval two-dimensional visual pattern that depends upon where one sits, or a set of tactile sensations that depend on where one's hand rests. In science we need "objective" descriptions of the experienced world. We need descriptions that do not depend on who it is that has the experience. Operational specifications fill this need. They are descriptions of possible human experiences that do not refer specifically to any particular individual. They allow us to create a science that is thoroughly objective, yet securely rooted in the realm of ideas and experience.

Question 2: In the author's article[21] on the S-matrix interpretation of quantum theory it was admitted that the pragmatic interpretation of quantum theory leaves unanswered deep metaphysical questions about the nature of the world itself. And it was noted that the apparent absence of unanalyzable entities in quantum theory suggests a "web" structure of nature that somewhat resembles the structure proposed by Whitehead. Does the absence of similar remarks in the present work signify a retraction of the earlier views?

Reply: The aim of the present work is to describe the Copenhagen interpretation. More precisely, the aim is to describe this author's understanding of the essential common ground of Bohr and Heisenberg on the question of the interpretation of quantum theory. The author's own views are an elaboration upon his understanding of the Copenhagen interpretation, and are given in the S-matrix article.

APPENDIX B: CORRESPONDENCE WITH HEISENBERG AND ROSENFELD

The views that have been put forth as representations of the Copenhagen interpretation differ widely. Thus the question arises whether my description succeeds in capturing the essence of the Copenhagen interpretation as understood by Bohr and Heisenberg. To shed light on this question I inquired of Heisenberg whether the description given in a first version of this paper seemed to him basically in accord with the views of himself and Bohr, or whether it seemed different in any important way.

Heisenberg replied:

> Many thanks for your letter and for your paper on the Copenhagen interpretation. I think that your text is a basically adequate description of the Copenhagen interpretation, and you probably know that Niels Bohr was very interested in the ideas of William James. I would, however, like to mention one point where you seem to describe the Copenhagen interpretation too rigorously. On p. 35 you ask the question "Can any theoretical construction give us testable pre-

dictions about physical phenomena that cannot be extracted from a quantum theoretical description?" and you say that according to the Copenhagen interpretation no such construction is possible. I doubt whether this is correct with respect to, for example, biological questions. Logically it may be that the difference between the two statements: "The cell is alive" or "The cell is dead" cannot be replaced by a quantum theoretical statement about the state (certainly a mixture of many states) of the system. The Copenhagen interpretation is independent of the decision of this point. It only states that an addition of parameters in the sense of classical physics would be useless.

Besides that it may be a point in the Copenhagen interpretation that its language has a certain degree of vagueness, and I doubt whether it can become clearer by trying to avoid this vagueness.

The paper was revised so as to make it absolutely clear that the claim of completeness of quantum theory refers specifically to atomic phenomena. Some superfluous material was eliminated, and the present Secs. V and VII, with their extensive quotations from Bohr, were added. Heisenberg's comments on the revised version were as follows:

> Many thanks for sending me the new version of your paper on the Copenhagen interpretation. It is certainly an improvement that you quote Bohr extensively, and your whole paper has become more compact and more understandable after these changes. There is one problem which I would like to mention, not in order to criticize the wording of your paper, but for inducing you to more investigation of this special point, which is however a very deep and old philosophical problem. When you speak about the ideas (especially in Chap. IV), you always speak about the human ideas, and the question arises, do these ideas "exist" outside of the human mind or only in the human mind? In other words: Have these ideas existed at the time when no human mind existed in the world?
>
> I am enclosing the English translation of a passage in one of my lectures in which I have tried to describe the philosophy of Plato with regard to this point. The English translation was done by an American philosopher who, as I think, uses the philosophical nomenclature correctly. Perhaps we could connect this Platonic idea with pragmatism by saying: It is "convenient" to consider the ideas as existing even outside of the human mind because otherwise it would be difficult to speak about the world before human minds have existed. But you see at these points we always get easily at the limitation of language, of concepts like "existing," "being," "idea," etc. I feel that you have still too much confidence in the language, but that you will probably find out yourself.

I replied:

> Regarding the question of nonhuman ideas it seems to me unlikely that human ideas could emerge from a universe devoid of idealike qualities. Thus I am inclined to the view that consciousness in some form must be a fundamental quality of the universe.... [However] It is difficult to extract from Bohr's writings any commitment on Platonic ideals. Indeed, Bohr seems to take pains to avoid all ontological commitment: He focuses rather on the question of how we as scientists are to cope with the limited validity of our classical intuitions.
>
> In view of Bohr's reluctance to speculate (in print at least) on the nature of the ultimate essences it has seemed to me that the consideration of these matters should not be considered a proper part of the Copenhagen interpretation. If the Copenhagen interpretation is considered to be an over-all world view that coincides with the complete world views of both you and Bohr, then there is danger that the Copenhagen interpretation may not exist; for it is not clear (from your respective writings at least) whether you and Bohr are in complete agreement on all ontological and metaphysical questions. Moreover, in your work *Physics and Philosophy* you discuss many of these

deeper philosophical questions, yet have a separate chapter entitled "The Copenhagen Interpretation." This suggests that "The Copenhagen Interpretation" should be interpreted in a restricted way. I have interpreted it to be not the complete over-all joint world view of Bohr and yourself, but rather the essential *common ground* of you and Bohr on the specific question of how quantum theory should be interpreted.

My practical or pragmatic account of quantum theory was based on the account given in your chapter "The Copenhagen Interpretation." This concrete account jibes completely with the abstract pronouncements of Bohr, as the quotations of Bohr in my Sec. V bear witness. Thus I think it correct and proper to regard the pragmatic interpretation of the formalism as an integral part of the Copenhagen interpretation. Similarly, I drew from our conversations at Munich an understanding of your commitment to the position that quantum theory provides for a complete description of atomic phenomena, and this position seems completely in accord with that of Bohr. Thus I think it correct and proper to regard also this position as an essential part of the Copenhagen interpretation. But in view of Bohr's silence on Platonic ideals I would hesitate to include considerations on that question in my account of the "logical essence of the Copenhagen interpretation." This is not meant to suggest that the Copenhagen interpretation bans further search for a comprehensive world view. It indicates only that the Copenhagen interpretation is, in my view, not itself a complete over-all world view: It is merely part of an over-all world view; the part that establishes the proper perspective on quantum theory. I emphasized in the closing passages of my paper that man's search for a comprehensive world view is not terminated by the Copenhagen interpretation. Rather it is significantly advanced.

Heisenberg replied:

Many thanks for your letter. May I just briefly answer the relevant questions. I agree completely with your view that the Copenhagen interpretation is not itself a complete over-all world view. It was never intended to be such a view. I also agree with you that Bohr and I have probably not looked upon the Platonic ideals in exactly the same way, and therefore there is no reason why you should go more into the problems of the Platonic ideals in your paper. Still there is one reservation which I have to make in connection with your paper and which I mentioned in my last letter. I think that you have too much confidence in the possibilities of language. I think that the attitude which is behind the Copenhagen interpretation is not compatible with the philosophy of Wittgenstein in the *Tractatus*. It may be compatible with the philosophy contained in the later papers of Wittgenstein. As you probably know, Bertrand Russell liked the *Tractatus* of Wittgenstein, but disapproved of the later papers and therefore I could never come to an agreement with Russell on these philosophical questions.

I replied:

Thank you for your very informative letter. I had not previously fully appreciated the point you were making, which as I now understand it, is this: You regard recognition of imperfectability of language to be an important element of the attitude that lies behind the Copenhagen interpretation. This point was not brought into my account of the Copenhagen interpretation, and is indeed somewhat at odds with its avowed aim of clarity. . . . [But] scientists must strive for clarity and shared understandings, since without striving even the attainable will not be achieved. . . .

Your words on the matters raised in our correspondence would certainly be extremely valuable to readers of my article. And any paraphrasing I might make would diminish this value. Thus, with your approval, I would like to include the full content of your letters (apart from personal openings and closings) in an appendix to my paper, along with certain connecting excerpts from my own letters. I have enclosed a copy of the proposed appendix, apart from your reply to the present letter.

Heisenberg replied: "Many thanks for your letter. I agree with your intention to publish my letters in the appendix to your paper."

I required also of Rosenfeld, as the close companion and co-worker of Bohr, and prime defender of his views, for an opinion of the extent to which my description succeeded in capturing the essence of the Copenhagen interpretation as it was understood by Bohr.

Rosenfeld expressed full agreement with my account, and gave hearty approval. He went on to comment on the relationship of Bohr to James. I include his remarks because of their historical interest:

> It may interest you to know that I several times endeavoured to persuade Bohr to make explicit mention of the affinity between his attitude and that of James, but he firmly refused to do so; not because he disagreed, but because he intensely disliked the idea of having a label stuck onto him. Indeed you may have noticed that some philosophers are already busy tracing imaginary influences of various philosophers upon Bohr. With regard to William James, I am quite sure that Bohr only heard of him from his friend, the psychologist Rubin, and from myself in the '30's. He then expressed enthusiastic approval of James' attitude, which he certainly felt akin to his own; but it is a fact—a very significant one, I think,—that James and Bohr developed a pragmatic epistemology independently of each other.
>
> It might be advisable to add somewhere in your paper a remark to that effect in order to avoid further misunderstanding. As a matter of fact, I have never myself in the papers I wrote on complementarity brought the pragmatic aspect of Bohr's thinking in explicit relation with James, precisely in order to avoid such misunderstanding.

He went on to say:

> I notice from your further letters with new title pages that you hesitate about the best title for your essay. I have no very strong view about this, but I would incline to prefer your March 31 title ["Quantum Theory, Pragmatism, and the Nature of Space–time"], the reason being that it does not contain the phrase "Copenhagen interpretation," which we in Copenhagen do not like at all. Indeed, this expression was invented, and is used by people wishing to suggest that there may be other interpretations of the Schrödinger equation, namely their own muddled ones. Moreover, as you yourself point out, the same people apply this designation to the wildest misrepresentations of the situation. Perhaps a way out of this semantic difficulty would be for you to say, after having pointed out what the difficulty is, that you make use of the phrase "Copenhagen interpretation" in the uniquely defined sense in which it is understood by all physicists who make a correct use of quantum mechanics. Surely, this is a pragmatic definition.

* This work was supported by the U. S. Atomic Energy Commission.

1 K. R. Popper and M. Bunge, in *Quantum Theory and Reality*, edited by M. Bunge (Springer, New York, 1967).

2 L. E. Ballentine, Rev. Mod. Phys. **42**, 358 (1970).

3 *Quantum Theory and Beyond*, edited by E. Bastin (Cambridge U. P., Cambridge, Eng., 1970); L. Rosenfeld, Suppl. Progr. Theoret. Phys. (Extra number) **222** (1965); J. M. Jauch, E. P. Wigner, and M. M. Yanase, Nuovo Cimento **48B**, 144 (1967); L. Rosenfeld, Nucl. Phys. **A108**, 241 (1968); A. Loinger, Nucl. Phys. **A108**, 245 (1968); Arthur Fine, Phys. Rev. **D2**, 2783 (1970).

4 Niels Bohr, *Atomic Theory and the Description of Nature* (Cambridge U. P., Cambridge, Eng., 1934); *Atomic Physics and Human Knowledge* (Wiley, New York, 1958); *Essays 1958/1962 on Atomic Physics and Human Knowledge* (Wiley, New York, 1963).

5 W. Heisenberg, *The Physical Principles of the Quantum Theory* (Dover, New York, 1930); in *Niels Bohr and the Development of Physics*, edited by W. Pauli (McGraw–Hill, New York, 1955), p. 12; *Physics and Philosophy* (Harper and Row, New York, 1958); Daedalus **87**, 95 (1958).

6 C. F. von Weiszäcker in Ref. 3.

7 J. von Neumann, *Mathematical Foundations of Quantum Mechanics* (Princeton, U. P., Princeton, N. J., 1955).

8 E. Wigner, Amer. J. Phys. **31**, 6 (1963).

9 G. Ludwig in *Werner Heisenberg und der Physik unserer Zeit* (Friedrich Vieweg, Braunschweig, 1961).

10 E. Wigner in *The Scientist Speculates*, edited by I. J. Good (Basic Books, New York, 1962), p. 284.

11 Hugh Everett III, Rev. Mod. Phys. **29**, 454 (1957).

12 J. A. Wheeler, Rev. Mod. Phys. **29**, 463 (1957).

13 Bryce DeWitt, in Phys. Today **23**, 30 (Sept. 1970).

14 L. Rosenfeld, Nucl. Phys. **A108**, 241 (1968).

AJP Volume 40

[15] D. Bohm, Phys. Rev. **85**, 166, 180 (1952).

[16] J. S. Bell, Physics (N. Y.) **1**, 195 (1964) and Varenna Lectures, Preprint Ref. TH.1220-CERN, Aug. 1970. See also Ref. 21.

[17] William James, *The Meaning of Truth* (Univ. Michigan, Ann Arbor, Mich., 1970). This reference to James does not mean that the ideas reviewed in this section are exactly those of James or wholly those of James. Countless philosophers have said similar things.

[18] Reference 17, p. 239.

[19] Reference 17, p. 217.

[20] A. Einstein, in *Albert Einstein, Philosopher–Physicist*, edited by P. A. Schlipp (Tudor, New York, 1951), p. 675.

[21] H. P. Stapp, Phys. Rev. **D3**, 1303 (1971).

[22] N. Bohr, Phys. Rev. **48**, 696 (1935) and in Ref. 20, p. 201.

[23] A. Einstein, in Ref. 20, p. 665.

[24] For an interesting and very readable account of the four principal conceptual structures that have been advanced as the basis of overall world views see S. C. Pepper, *World Hypothesis* (U. of California Press, Berkeley, Calif., 1966).

IL NUOVO CIMENTO VOL. 77 B, N. 1 11 Settembre 1983

Reprinted from *Nuovo Cimento,* **77B**, M. Bunge and A.J. Kalnay, "Solution to Two Paradoxes in the Quantum Theory of Unstable Systems," pp. 1-9, ©1983.

Solution to Two Paradoxes in the Quantum Theory of Unstable Systems.

M. BUNGE

Foundations and Philosophy of Science Unit
McGill University - Montreal, P.Q., Canada H3A 1W7
Institute for Basic Research - Cambridge, MA 02138

A. J. KÁLNAY

Centro de Física, Instituto Venezolano de Investigaciones Científicas (IVIC)
Apdo. 1827, Caracas 1010 A, Venezuela
Institute for Basic Research - Cambridge, MA 02138

(ricevuto il 14 Marzo 1983; manoscritto revisionato ricevuto il 20 Giugno 1983)

Summary. — Two apparent paradoxes in the quantum theory of unstable systems are discussed and dissolved. According to one of them all decay processes, regardless of their mechanism, would be exponential under certain very general assumptions. According to the other an unstable system will never decay, provided it is kept under constant observation. The paradoxes are shown to derive from the tacit classical assumption that every unstable thing is always either in the undecayed or in the decayed state. They do not occur as long as the superposition principle is taken seriously and the system is conceived as being normally in a superposition of sharp states. Moreover, the decay is shown to be a continuous process that unfolds regardless of the observations to which the decay products may be subjected.

PACS. 03.65. – Quantum theory; quantum mechanics.

1. – Introduction.

Although radioactivity was a major source of quantum theory, the vast literature on unstable systems is somewhat confusing and even ridden with paradox. (See (1) for a masterly survey of the entire field and (2-4) for incisive analyses of some remarkable paradoxes.) Some authors find it puzzling that not all decay processes are strictly exponential. Two authors, while claiming to prove that all decay processes are exponential regardless of their mechanism, acknowledge that this view « is incompatible with the accepted physical interpretation of quantum mechanics » (5). Still others wonder whether the rate of decay depends critically upon observation (6). Still others (2-4) claim that quantum mechanics may involve a sort of Zeno's paradox, in that an unstable system subject to continual observation would not change, let alone decay. Whereas some physicists have accepted this last statement, others have tried to get around it by invoking epistemological arguments concerning the nature of observation, and still others have suggested that quantum mechanics is just incapable of computing certain transition probabilities (2), so it may have to be subjected to a fundamental revision.

The aim of the present paper is to trace the sources of these puzzles. Its upshot is that they concern change in general, not just the decay of unstable systems, and that they can all be solved without modifying the mathematical formalism of quantum theory. It turns out that what must be given up instead are *a*) the classical presupposition that every dynamical variable has a definite (sharp) value at every time, *b*) the idealization of the instantaneous measurement (that takes no time) (7) and *c*) the subjectivistic thesis that the observer is essential for projecting the state vector of the system onto an axis (or, more generally, a subspace) of the Hilbert space of the latter—in particular on the initial undecayed state. Our results are obtained by *a*) sticking to the superposition principle, *b*) assuming that each step in a sequence of measurements takes a nonvanishing time interval (7) and *c*) abiding by an objectivistic interpretation of the quantum theory of a single system—though, of course, a probabilistic one.

2. – Two paradoxes.

Let us start by making a quick review of two typical and alarming paradoxes occurring in the theory of unstable systems and, indeed, of any changing things. According to one of them random decay processes are of the expo-

---

(1) L. Fonda, G. C. Ghirardi and A. Rimini: *Rep. Prog. Phys.*, **41**, 587 (1978).
(2) B. Misra and E. C. G. Sudarshan: *J. Math. Phys.* (*N.Y.*), **18**, 756 (1977).
(3) C. B. Chiu, E. C. G. Sudarshan and B. Misra: *Phys. Rev. D*, **16**, 520 (1977).
(4) A. Peres: *Am. J. Phys.*, **48**, 931 (1980).
(5) H. Ekstein and A. J. F. Siegert: *Ann. Phys.* (*N. Y.*), **68**, 509 (1971).
(6) A. Degasperis, L. Fonda and G. C. Ghirardi: *Nuovo Cimento A*, **21**, 471 (1974).
(7) This point will be taken up in the following paper by the same authors, *Nuovo Cimento B*, **77**, 10 (1983).

nential type provided decay observations at intermediate times leave the decay probability unchanged. According to the second paradox continual observation ensures perpetuity. Both paradoxes are independent of the nature of the components of the system as well as of the interactions among them: *i.e.* the particular form of the Hamiltonian of the system is irrelevant. Let us begin with the first paradox.

Consider an unstable system that can be in either of two states, to be called the initial (or excited or undecayed) and the final (or ground or decayed) states. (Pay attention to the exclusive disjunction « either ».) Assume that the probability of the transition from the initial to the final state depends only upon time. We proceed to recall the proof that, under the preceding assumptions, the probability that the system stays in its initial state by time t equals $\exp[-\lambda t]$, where $\lambda > 0$ is a parameter characterizing the kind of system. (This is, of course, just a special case of the Poisson distribution; for the general case see, *e.g.*, ([8]).) The proof is worth watching for the role that a certain assumption, usually tacit, plays in it.

Call $P(t')$ the probability that the system has remained in its initial state up until time t' and $P(t'')$ the probability that it has not decayed by time t''. (If preferred, reason on a population of similar systems and consider the average number of undecayed systems, obtained by multiplying the probabilities by the number of systems present at the initial time.) Since by hypothesis the decay probability depends only upon the time elapsed since the initial time $t = 0$, having stayed in the initial state by time t' and not having changed state between t' and t'' are statistically independent events. Thus $P(t'')$ equals the product of $P(t')$, the probability that nothing has altered up until t', and $P(t'' - t')$, the probability that nothing changes during the subsequent time interval. In short,

$$P(t')\cdot P(t''-t') = P(t'') ,$$

or, in a different notation,

$$P(t_1)\cdot P(t_2) = P(t_1 + t_2) . \tag{1}$$

The only continuous solution to this functional equation is the exponential function a^{bt}, where a and b are arbitrary real numbers. Because probabilities are bounded between 0 and 1, we may choose $a = e$ and must choose a negative value b: call it $-\lambda$ with $\lambda > 0$. Thus finally we get

$$P(t) = \exp[-\lambda t] . \tag{2}$$

([8]) W. Feller: *An Introduction to Probability Theory and Its Applications*, Vol. 1, 3rd edition (New York, N. Y., 1968).

This result is supposed to be quite general: it should hold for an arbitrary two-state system regardless of its dynamics, provided its decay probability depends only upon time, and that the system be at any time in either of the two states rather than in a superposition of them—but this latter condition does not hold in quantum mechanics. Here only stable (stationary) states are eigenstates of the energy operator. If the system is inherently unstable (rather than as a result of external perturbations), it cannot be in an eigenstate of its Hamiltonian; correspondingly its energy, far from being sharp, will be spread over a nonvanishing interval.

Consequently, when calculating the probability that the system be in its undecayed state at a given instant, we must take into account all the possible transitions among the states that compose the wave packet in momentum space. By so doing we shall be reasoning on state vectors rather than on probabilities, which was how eq. (2) was arrived at. (Recall the wrong results concerning the double-slit experiment when reasoning directly on probabilities: we lose the so-called interference terms.) However, so strong are the hold of eq. (2) and of the underlying assumption that the system must always be either in its initial state or in its decayed state that any quantum-theoretical deviations from eq. (2)—*e.g.* for very small or very large times—sound paradoxical. (See [1] for such deviations, in particular for the result that the decay function has a vanishing derivative at the origin.)

The second paradox too can be constructed with reference to a two-state system. Call φ the initial (excited) state of the system and $\hat{H}$ its Hamiltonian operator. The probability that the system be in its excited state at time t equals the square of the projection of its state vector at time t onto the initial state, *i.e.*

$$q(t|0) = |(\varphi, \exp[-i\hat{H}t/\hbar]\,\varphi)|^2 . \tag{3}$$

This is the conditional probability that the system has either remained in its initial state or returned to it any number of times by time t. Expanding the evolution operator as a Taylor series and keeping only the first two terms, we obtain a result valid for small values of t:

$$q(t|0) \simeq 1 - (\sigma/\hbar)^2 t^2 , \tag{4}$$

where σ, often called ΔE, is the mean standard deviation of $\hat{H}$.

Normally, to obtain the value of $q(t|0)$ at any instant (in not too distant a future), one sets t equal to the desired value on the r.h.s. of eq. (4). Suppose, however, we reason along a line similar to the one that led to the preceding paradox. Consider the unit time interval $[0, 1]$ and divide it into n equal subintervals. The probability that the system is in its initial state at the end of the first interval $1/n$ is, by eq. (4), $q_1 \simeq 1 - (\sigma/n\hbar)^2$. Since by eq. (3) the decay probability does not depend upon the previous history of the system (statistical independence), after the second interval has elapsed we shall have $q_2 = [1 - (\sigma/n\hbar)^2]^2$. Finally, after the unit interval n/n, we get

$$q_n = [1 - (\sigma/n\hbar)^2]^n , \tag{5}$$

which approaches 1 as n approaches infinity. That is, the system stays in its initial state regardless of the length of the time interval. This paradox, attributed to TURING (9), has come to be known as Zeno's quantum paradox (2-4).

This result is usually interpreted in terms of measurement operations (2-4). It is said that the system is being observed at the end of each subinterval m/n for $1 \leqslant m \leqslant n$. Every time an observation is made, the state vector of the system would collapse onto its initial state φ in obedience to von Neumann's projection postulate. If the observations are made continuously, the system is prevented from changing because its state vector is forced all the time to lie down on the initial state. Thus the quantum theory of measurement would supply at low cost the Fountain of Youth that PONCE DE LEÓN sought in vain at great peril. We intend to show that this is not the case.

3. – Solution to the paradoxes.

Consider an unstable system in the absence of any macrophysical system capable of interfering with it. Think for definiteness of a neutron ejected into a vacuum by a nuclear reaction. The neutron may eventually decay according to the scheme $\mathrm{n} \to \mathrm{p} + \mathrm{e} + \bar{\nu}_e$. As before, call φ the initial state of the neutron and

$$\psi(t) = \exp[-i\hat{H}t/\hbar]\,\varphi \tag{6}$$

its state at any later time t. This state vector is a linear combination of φ with states orthogonal to φ, except in the particular case when φ happens to be an eigenstate of $\hat{H}$, in which case there is no disintegration. In other words, no sooner was the neutron formed, than it had a propensity (objective probability) of continuing to be a neutron and another of turning into a proton,

(9) As reported by W. YOURGRAU in I. LAKATOS and A. MUSGRAVE, Editors: *Problems in the Philosophy of Science* (Amsterdam, 1968), p. 191.

an electron and an antineutrino. (For the notion of objective probability see ([10-15]).)

In other words, unless the neutron has disintegrated by time t, it is *neither* in the sharp n state *nor* in the sharp (p, e, $\bar{\nu}_e$) state, but in a linear combination of the two. Hence, the usual view that the neutron is a neutron, as long as it has not disintegrated, is incorrect: neutrons and other quantum objects are fuzzy rather than definite—save in exceptional circumstances, as when environmental (*e.g.* experimental) circumstances force them to go into some sharp state. This is the clue to our dissolution of the two paradoxes discussed in sect. **2.**

The paradoxes come from assuming that the system is at all times *either* in its initial (undecayed) state *or* in its final (decayed) state. Thus in the first case we obtain eq. (1) and its consequences, which are not generally true. In the second case we obtain eq. (5) and its paradoxical consequences. The gist of this conundrum is as follows.

Call now φ_1 the initial (undecayed) state and φ_2 the final (decayed) state of a system which, like the free neutron of the above example, is not being tampered with by any macrosystem. (We shall deal with experimental situations in ref. ([7]).) According to quantum theory, at any time $t > 0$ the system is normally in a superposition of states:

$$\psi(t) = c_1(t)\varphi_1 + c_2(t)\varphi_2 \,, \tag{7}$$

where c_1 and c_2 are the projections of ψ onto the initial and the final states, respectively, *i.e.*

$$c_1(t) = (\varphi_1, \psi(t)) \,, \qquad c_2(t) = (\varphi_2, \psi(t)) \,, \tag{8a}$$

where

$$|c_1(t)|^2 + |c_2(t)|^2 = 1 \qquad \text{for all } t. \tag{8b}$$

The physical meaning of these projections is well known:

$|c_1(t)|^2 =$ probability that the system be in state φ_1 at time t,

$|c_2(t)|^2 =$ probability that the system be in state φ_2 at time t.

([10]) M. v. Smoluchowski: *Naturwissenschaften*, **6**, 253 (1918).
([11]) M. Fréchet: *Les mathématiques et le concret* (Paris, 1946), p. 157.
([12]) M. Bunge: *Br. J. Philos. Sci.*, **6**, 1, 141 (1955).
([13]) M. Bunge: *Am. J. Phys.*, **24**, 272 (1956).
([14]) K. R. Popper: *Observation and Interpretation*, edited by S. Körner (London, 1957).
([15]) M. Bunge: *Appl. Math. Model.*, **5**, 306 (1981).

Equations (8) suggest a simple geometrical representation that brings out the profound differences between the quantum and the classical theories. Assuming that $c_1(t)$ and $c_2(t)$ are real numbers, rewrite them, without loss of generality, as

$$c_1(t) = \cos\alpha(t)\,, \qquad c_2(t) = \sin\alpha(t) \qquad \text{with} \qquad \alpha(0) = 0\,. \tag{9}$$

For small values of t we may use eq. (4), setting

$$\alpha(t) = \omega t \qquad \text{with} \qquad \omega \simeq \sigma/\hbar\,, \tag{10}$$

so that

$$c_1^2 = \cos^2\omega t \simeq 1 - (\sigma/\hbar)^2 t^2\,, \qquad c_2^2 = \sin^2\omega t \simeq (\sigma/\hbar)^2 t^2\,. \tag{11}$$

The physical interpretation of the preceding equations is obvious. As time goes by, the state vector ψ rotates with angular velocity ω about the origin of the state space formed by the axes φ_1 and φ_2. In other words, the system goes back and forth between the two states every $\hbar/\sigma$ seconds. In the classical case there is no rotation because the energy is sharp at all times, *i.e.* $\sigma = 0$. In this case the state vector ψ lies always on the initial state φ_1: nothing happens. In particular, there is no quantum jump from state φ_1 to state φ_2. (Nor is there a jump proper in the quantum-theoretical case!)

The paradox dissolves then upon taking the superposition principle (eq. (7)) seriously. In other words, there is no paradox if the system is not regarded as being at every time in an eigenstate of its energy operator. A treatment of the unstable system subjected to repeated observations will be given in ref. (7).

4. – Concluding remarks.

The two paradoxes encountered in the theory of unstable systems can be traced back to the principle that every property of a physical system has a definite (sharp, well defined) value at any given time—or, equivalently, that the system is in a particular state of some dynamical variable, such as the energy, at any time. (In the case of the neutron, discussed in sect. **3**, this dynamical variable is the 3rd component of the isospin.) This principle, sometimes wrongly dubbed «realist», is inherent in classical physics, but it does not hold in quantum theory.

As BOHR emphasized repeatedly (16), the superposition principle is central to quantum theory. It is more than a mathematical theorem holding for all

(16) N. BOHR: *Atomic Theory and the Description of Nature* (Cambridge, 1934).

linear theories and allowing one to expand an arbitrary state vector in eigenvectors of an arbitrary dynamical variable. Indeed, the superposition principle states also that a physical entity is normally in a superposition of sharp states—and this quite independently of considerations of measurement. Abiding by this principle and abstaining from invoking any magical actions of the allegedly omnipotent observer suffices to dissolve the paradoxes of the exponential decay of all unstable systems and of the perpetuity of unstable systems subjected to continual observation.

Consequently we do not subscribe to the pessimistic view ([2]) that quantum theory lacks a trustworthy algorithm for computing decay probabilities and must, therefore, be suspected of being incomplete in a serious regard. (There are, of course, problems with quantum theory, as with every other rich scientific theory, but they lie elsewhere ([17,18]).) The following paper (ref. ([7])), devoted to the observation of an unstable system in a realistic sequence of nonvanishing time periods, should confirm our verdict: *No paradox this time.*

If preferred, what may seem paradoxical instead are two results that lie buried in sect. **3**. One is that, strictly speaking, there can be no such thing as a two-state unstable quantum system. The reason is that every unstable state must be construed as a superposition of infinitely many sharp states. The other seeming paradox is that, precisely for this reason, there are no quantum jumps in quantum theory, where the state vector evolves smoothly according to the Schrödinger equation. True, the collapse or reduction of the wave function upon measurement is usually regarded as a discontinuous event. However, the reduction may be conceived of as a very swift process satisfying the Schrödinger equation, and it will be so conceived in ref. ([7]). Besides, the reduction to a single eigenstate happens only in the case of discrete variables such as electric charge and spin; for continuous ones the original wave packet must be thought of as collapsing onto a smaller bundle of states (*i.e.* an eigendifferential) ([19]).

* * *

One of us (MB) is grateful to the Social Sciences and Humanities Research Council of Canada for supporting his research. AJK thanks the IVIC for a travel grant that has allowed him to enjoy the hospitality of the Foundations and Philosophy of Science Unit.

([17]) M. BUNGE: *Foundations of Physics* (New York, N. Y., 1967).

([18]) P. A. M. DIRAC: *The Physicist's Conception of Nature*, edited by J. MEHRA (Dordrecht, 1973), p. 1.

([19]) *Footnote added in proofs.* – This view, that the reduction of the wave packet *a*) is generally only approximate, *b*) does not require an observer and *c*) is a consequence of the Schrödinger equation, is, of course, at variance with the usual interpretation of quantum mechanics. See M. CINI: *Nuovo Cimento B*, **73**, 27 (1983) for a forceful argument in favour of that unorthodox view.

● RIASSUNTO (*)

Si discutono e risolvono due apparenti paradossi nella teoria quantistica di sistemi instabili. Secondo uno di questi, tutti i processi di decadimento, indipendentemente dal loro meccanismo, sarebbero esponenziali secondo certe ipotesi molto generali. Secondo l'altro, un sistema instabile non decaderà mai, ammesso che sia mantenuto sotto costante osservazione. Si mostra che i paradossi derivano dal sottinteso presupposto classico che ogni cosa instabile sia sempre sia nello stato non decaduto che in quello decaduto. Questi non si verificano fino a che il principio di sovrapposizione è preso seriamente e il sistema è considerato essere normalmente in una sovrapposizione di stati definiti. Inoltre si mostra che il decadimento è un processo continuo che si svolge indipendentemente dalle osservazioni alle quali i prodotti di decadimento possono essere soggette.

(*) *Traduzione a cura della Redazione.*

Решение двух парадоксов в квантовой теории нестабильных систем.

Резюме (*). — Обсуждаются и разрешаются два кажущихся парадокса в квантовой теории нестабильных систем. Согласно одному из них, все процессы распада, независимо от их механизма, будут экспоненциальными при некоторых очень общих предположениях. Согласно другому, нестабильная система никогда не распадется, при условии, что она находится под постоянным наблюдением. Показывается, что парадоксы возникают из-за неявного классического допущения, что каждый нестабильный объект всегда находится либо в нераспавшемся, либо в распавшемся состояних. Парадоксы не возникают, пока имеет место принцип суперпозиции и система представляется, как суперпозиция резких состояний. Более того, показано, что распад является непрерывным процессом, который развертывается независимо от наблюдения, которое может воздействовать на продукты распада.

(*) *Переведено редакцией.*

M. BUNGE, *et al.*
11 Settembre 1983
Il Nuovo Cimento
Serie 11, Vol. **77** B, pag. 1-9

PHYSICS TODAY

VOL 22 NO 4

AN OPERATIONAL INTERPRETATION OF

NONRELATIVISTIC QUANTUM MECHANICS

Idealized experiments show that a physical measurement involves three stages: initial state preparation, perturbation and examination for the probability that a particular final state is occupied.

WILLIS E. LAMB JR

WHAT IS QUANTUM MECHANICS? A remarkable feature of the 1968 conference of Nobel prize winners in physics at Lindau is that it was possible for me to ask such a question in the presence of two of the founders of quantum mechanics, Werner Heisenberg and P.A.M. Dirac, more than 30 years after the discovery, in a lecture attended by 400 students who had recently begun their study of the subject. Several answers to the question are possible. The only easy one is that quantum mechanics is a discipline that provides a wonderful set of rules for calculating physical properties of matter. For such simple systems as hydrogen and helium atoms the calculated energy levels agree with experiment to fantastic accuracy. In more complicated cases the computations are difficult and the accuracy is lower, but it is reasonable to believe, in principle at least, that the theory would be adequate if only the calculational problems could be overcome.

A discussion of the interpretation of quantum mechanics on any level beyond this almost inevitably becomes rather vague. The major difficulty involves the concept of "measurement," which in quantum mechanics means determining the value of a physical observable for a dynamical system with as much precision as is possible.

I have taught graduate courses in quantum mechanics for over 20 years at Columbia, Stanford, Oxford and Yale, and for almost all of them have dealt with measurement in the following manner. On beginning the lectures I told the students, "You must first learn the rules of calculation in quantum mechanics, and then I will tell you about the theory of measurement and discuss the meaning of the subject." Almost invariably, the time allotted to the course ran out before I had to fulfill my promise.

The truth was that I found most available discussions of quantum-mechanical measurement either too vague, or too formal and unphysical. However, as the years went by, I gradually put together for myself an interpretation of nonrelativistic quantum mechanics that stresses an operational point of view. My attitude toward such problems has no doubt been influenced by contact with some research in experimental physics in which atomic states are "manipulated" by microwave and radiofrequency fields. In discussion of the measurement of some dynamical variable of a physical system I want to know exactly what apparatus is necessary for the task and how to use it, at least in principle. I am not satisfied with hand waving or with a formal logical scheme involving black boxes.

Mark Twain once said about meteorology: "Everybody talks about the weather, but nobody does anything about it." A similar statement could be made about the subject of measurement in quantum mechanics. Whereas no one, in fact, could actually carry out any of the experimental procedures I will be discussing, they represent an ideal limit that is worth considering. Granting reasonable premises, one knows, as far as possible, at all times, exactly what one is doing to the physical system under consideration. If, in practice, one is less than certain about what he is doing to the system, then it appears obvious to me that his knowledge of its future state will be reduced. I believe, in fact, that the measurements I will describe are the most precise ones possible. They involve the use of idealized apparatus that may not be realizable in practice. Actual measurements are less than ideal. In almost all physical experiments where one says that something is "measured" there is at most only a very bad quantum-mechanical measurement in the technical sense of the term. This is certainly true of the scattering experiments of atomic and nuclear physics. For example, the famous "gamma-ray microscope" (figure 1) used by Heisenberg[1] for qualitative discussions of his uncertainty relations does not provide a measurement of position but a scattering experiment. In my opinion, a quantum-mechanical theory of imperfect measurements, in which one settles for less than the best imaginable experimental control, has still to be given. It will probably make use of the theory of random processes.

Spinless electrons in a field

Let us consider that the physical system under study is a nonrelativistic

Willis E. Lamb Jr's Nobel Prize was awarded in 1955 for the discovery of what everyone now knows as the "Lamb shift" in the fine structure of the hydrogen spectrum. He served on the physics faculties of Columbia, Stanford and Oxford Universities before taking up his present position as Henry Ford II professor of physics at Yale in 1962.

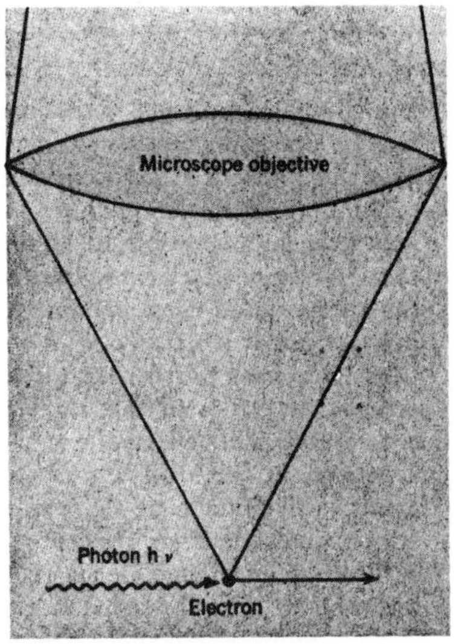

GAMMA-RAY MICROSCOPE postulated by Heisenberg in his discussions of the uncertainty relations. The electron cannot be observed unless it interacts with a photon, and then its recoil makes accurate simultaneous position and momentum measurements impossible. —FIG. 1

spinless electron of mass m moving in a known conservative force field $V(x)$. Its Hamiltonian operator is the sum of a kinetic and potential energy

$$H = T + V$$

with

$$T = -\frac{\hbar^2}{2m}\nabla^2, \qquad V = V(x)$$

Associated with the electron is a wave function $\psi(x,t)$ that obeys the Schrödinger wave equation

$$-\frac{\hbar}{i}\frac{\partial\psi}{\partial t} = (T + V)\,\psi(x,t)$$

This differential equation arises from an attempt to generalize classical mechanics by introducing a wave field $\psi(x,t)$ somehow describing the electron. According to the basic assumption of quantum mechanics

$$dW(x,t) = |\,\psi(x,t)\,|^2\,dx$$

gives the probability at time t that the electron is between x and $x + dx$. This probability statement refers not to a single system but to an ensemble of systems all of which are described by the same wave function $\psi(x,t)$. In further development of the theory, the probability assumption is generalized as follows: Consider a complete set of commuting Hermitian operators (observables) $F(x,p)$ with eigenvalues F_f and eigenfunctions $\psi_f(x)$ obeying the equation

$$F(x,p)\,\phi_f(x) = F_f\phi_f(x)$$

The functions $\phi_f(x)$ form a complete set of orthonormal expansion functions for the problem so that we may expand the wave function in the form

$$\psi(x,t) = \Sigma_f C_f(t)\,\phi_f(x)$$

It then follows plausibly that the absolute square of the expansion coefficient $C_f(t)$ is the probability

$$W_f = |C_f|^2$$

that the dynamical variables F "have" the values F_f for the state $\psi(x,t)$. Note that I am being rather vague about how the values F_f are to be determined. The probabilistic assumption carries with it no recipe for its experimental verification or rejection. It is also not clear whether all Hermitian operators F, or only some specially favored ones, can be measured in the sense implied by the probability assumption.

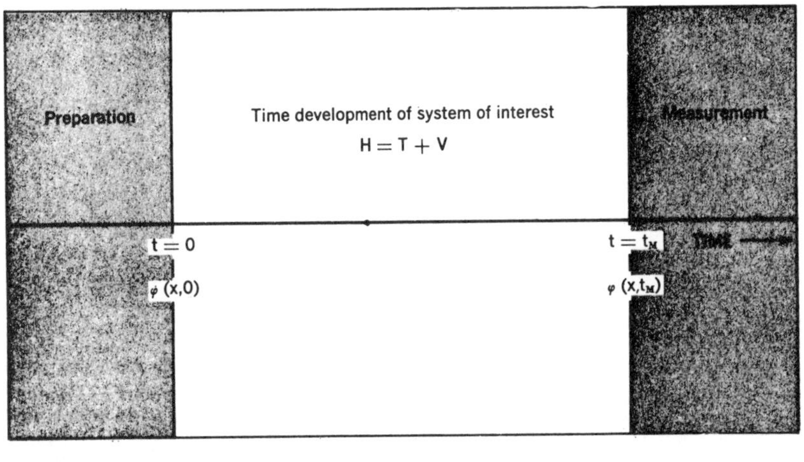

STATE PREPARATION AND MEASUREMENT. The wave function of a system develops from time $t = 0$ to time $t = t_M$ according to Schrodinger's time-dependent wave equation. The process of "preparation" of the initial state occurs before $t = 0$. A measurement of some physical quantity is begun at time t_M. —FIG. 2

Classical state preparation

Before indicating how measurements are to be made, I take up an important related question, the *preparation* of a state $\psi(x,0)$ at time $t = 0$. (Although some authors confuse *preparation of a state* and *measurement*, these concepts are logically and physically very different.) It is clear that the time-dependent Schrödinger equation will determine the future wave function $\psi(x,t)$ if its initial form $\psi(x,0)$ is known. For this to be useful we should have some way to start the system off with the initial wave function so that we can derive some physical information from its calculable value $\psi(x,t_M)$ at a time t_M of measurement. This is indicated schematically in figure 2. Just before $t = 0$ certain procedures called preparation of the initial state are carried out. From $t = 0$ until $t = t_M$ the system's wave function develops according to its Schrödinger wave equation. Just after $t = t_M$, a set of procedures constituting the measurement of some dynamical variable $F(x,p)$ are carried out on the system.

The first question to answer is: Can one give an experimental procedure for bringing the electron initially into a state having an arbitrarily assigned wave function $\psi(x,0)$? The answer is "Yes," provided that certain assumptions are made. Consider the corresponding problem in classical mechanics where the particle has the Hamiltonian $H(x,p)$. The initial state of a particle is fully specified by its classical coördinate $x(0)$ and momentum $p(0)$, and the future state $x(t)$, $p(t)$ is determined uniquely by integrating the Hamiltonian equations of motion

$$\dot{x} = \frac{\partial H}{\partial p}, \qquad \dot{p} = -\frac{\partial H}{\partial x}$$

subject to the initial conditions at $t = 0$. In classical mechanics one would ordinarily consider the preparation of the initial state as trivial. At $t = 0$ one simply places the particle at $x(0)$ and gives it a momentum $p(0)$. This can be done entirely within the bounds of classical mechanics. We set up a potential well $U_1(x)$ having a minimum at $x = x(0)$. Then we somehow catch a particle in the potential well and allow it to settle down to rest at the minimum with the assistance of a little friction. It is then located at

$x(0)$ at $t = 0$. We next give the particle an impulsive force derived from a potential $U_2(x)\ \delta(t)$ where $\delta(t)$ is the Dirac delta function, sharply peaked at $t = 0$. This force can bring the particle momentum to $p(0)$ without changing its position appreciably. Just afterwards, the potential $V(x)$ corresponding to the problem of interest is turned on. As we want to be able to do this for arbitrary values of $x(0)$, $p(0)$ it seems reasonable to assume that we can apply arbitrary potentials of the form $U(x,t)$ to the system and that we can introduce a small amount of frictional force when needed.

Quantum-mechanical preparation

We now turn to the problem of state preparation in quantum mechanics. It may come as a surprise that this requires, at least in principle, exactly the same experimental facilities as the classical case needed—and no more.

Suppose we want to prepare a state $\psi(x,0)$. If this happens to be known to us as the lowest energy eigenfunction for a certain potential, all we have to do is to set up this potential, somehow catch an electron in it, and wait for radiative damping to bring the particle to the ground state. Then we suddenly turn off $U_1(x)$ and turn on the $V(x)$ for the problem of interest. (According to sudden-perturbation theory[2] a wave function is not immediately changed by a sudden change in potential if no singular time dependence is involved.) The wave function $\psi(x,t)$ for later times would then be determined by the time-dependent Schrödinger equation

$$-\frac{\hbar}{i}\frac{\partial\psi}{\partial t} = H\psi = (T+V)\psi$$

for the system of interest. There is, of course, the problem of recognition that we have caught an electron in $U_1(x)$, but I regard this as trivial in principle and do not discuss it here.

If the desired initial state is not the known lowest-energy eigenfunction of some potential $U_1(x)$ a number of procedures are possible. We may look for a potential $U_1(x)$ in which $\psi(x,0)$ is an eigenstate of some energy E, if not the lowest one. Because $\psi(x,0)$ obeys the stationary-state wave equation

$$[T + U_1(x)]\ \psi(x,0) = E\ \psi(x,0)$$

we can solve for the required potential $U_1(x)$ and find explicitly that

$$U_1(x) = E - [\psi(x,0)]^{-1}\ T\ \psi(x,0) \quad (1)$$

If this potential $U_1(x)$ is experimentally available, we catch the electron in it. The wave function will then be some linear combination of the eigenstates of the Hamiltonian $T + U_1$, one of which is the desired $\psi(x,0)$. There are several possibilities here. One is to carry out a state-selection process of the Stern–Gerlach type to collect the atom if its energy eigenvalue is E. If we get some other eigenvalue, we reject the atom and repeat the procedure until we get the desired one.

Complications

Three difficulties might occur: The potential $U_1(x)$ given by equation 1 might be singular if $\psi(x,0)$ has nodes; the energy spectrum might not be discrete so that state selection would be difficult, or the potential $U_1(x)$ might be a complex function because $\psi(x,0)$ is not a real function. These troubles can either be ignored, or overcome by methods such as the one I will discuss here only for the case that the function $U_1(x)$ given by equation 1 turns out to be complex.

In that case, we write the desired initial wave function in the form

$$\psi(x,0) = R(x)e^{iS(x)}$$

where $R(x)$ and $S(x)$ are real functions of position. Then we use a real potential

$$U_1(x) = E - [R(x)]^{-1}\ T\ R(x) \quad (2)$$

instead of equation 1 and prepare a state with the real wave function $R(x)$ in the manner previously described. We next apply a pulse potential

$$U_2(x,t) = -\hbar\ S(x)\ \delta(t) \quad (3)$$

that is designed to convert the wave function from $R(x)$ to $\psi(x,0)$. During the pulse, Schrödinger's wave equation can be adequately approximated by

$$-\frac{\hbar}{i}\frac{\psi}{t} = -\hbar\ S(x)\ \delta(t)\ \psi(x,t)$$

or

$$\frac{\partial\psi(x,t)}{\partial t} = i\ S(x)\ \delta(t)\ \psi(x,t)$$

The solution of this differential equation is easily found. Just after the pulse the wave function is

$$\psi(x,0_+) = R(x)e^{iS(x)}$$

if initially we had the wave function $R(x)$. In this way, by first applying the potential $U_1(x)$ of equation 2, then $U_2(x,t)$ of equation 3 and finally $V(x)$, we can prepare any desired initial state of the system of interest. Note the close parallel with the classical situation.

Spreading wave packets

We might illustrate this procedure with a simple example. Consider a free particle whose Hamiltonian is

$$H = \frac{p^2}{2m}$$

and that has a real Gaussian wave function at $t = 0$,

$$\psi(x,0) = (2\pi)^{-1/4}(\Delta x_0)^{-1/2} \exp\left\{-\frac{1}{4}\frac{x^2}{(\Delta x_0)^2}\right\}$$

The corresponding probability density

$$\rho(x,0) = |\psi(x,0)|^2 = (2\pi)^{-1/2}(\Delta x_0)^{-1} \exp\left\{-\frac{1}{2}\frac{x^2}{(\Delta x_0)^2}\right\}$$

is also Gaussian and has a width $[\langle x^2\rangle - \langle x\rangle^2]^{1/2} = \Delta x_0$. According to the time-dependent wave equation, the wave function after elapse of a time T is the complex Gaussian expression[3]

$$\psi(x,T) = (2\pi)^{-1/4} \left[\Delta x_0 + \frac{1}{2}(i\hbar T)/(m\Delta x_0)\right]^{-1/2} \times \exp\left\{\frac{-x^2/4}{[(\Delta x_0)^2 + (1/2)\ (i\hbar T)/m]}\right\}$$

for which the probability density is

$$\rho(x,T) = |\psi(x,T)|^2 = (2\pi)^{-1/2}\left[(\Delta x_0)^2 + \frac{1}{4}(hT)^2/(m\Delta x_0)^2\right]^{-1/2} \times \exp\left\{\frac{-x^2/2}{[(\Delta x_0)^2 + (1/4)(hT)^2/(m\Delta x_0)^2]}\right\}$$

During the time T the wave packet has spread, remaining Gaussian, but its width parameter Δx increases with time according to the equation

$$(\Delta x)^2 = (\Delta x_0)^2 + \frac{\hbar^2T^2}{4m^2(\Delta x_0)^2}$$

This has an obvious interpretation in terms of Heisenberg's uncertainty relationship

$$\Delta x\,\Delta p \geq \frac{1}{2}\hbar$$

The initial spread Δx_0 in x implies an uncertainty in p, and this after a time T implies an additional spread in x equal to $\hbar T/(2m\Delta x_0)$, which has to be added in quadrature to the original Δx_0.

The above wave packet spreads for $t > 0$. If we could start at time T with a wave packet $\psi(x,T)$ and reverse the arrow of time, the width of the probability distribution would shrink from Δx at $t = T$ to Δx_0 at $t = 0$, whereas it would spread again for $t < 0$. Unfortunately, no one knows how to reverse the flow of time, although many theoretical papers ignore the difficulty. However, if we could start at time $t =$

0 with a wave packet having the initial form

$$\psi(x, -T) \equiv \psi(x,T)^*, T > 0$$

then the probability distribution would shrink from a width Δx at time $t = 0$ to Δx_0 at the later time $t = T$; then spreading would occur. There is no limit (in nonrelativistic quantum mechanics) to the sharpness Δx_0 of the wave packet that could thereby be achieved at time T. We must merely start with the proper complex wave packet $\psi(x, -T)$ with an enormous width parameter Δx. The question then arises: Can this wave packet be realized experimentally, and if so, how? Our general recipe for preparation of a state gives the answer. The complex wave function $\psi(x, -T)$ has to be written in the form

$$\psi(x, -T) = R(x)e^{iS(x)}$$

After a simple calculation we find that

$$R(x) = (2\pi)^{-1/4} (\Delta x)^{-1/2} \exp\left\{-\frac{1}{4}\frac{x^2}{(\Delta x)^2}\right\}$$

and apart from an additive constant

$$S(x) = -\frac{1}{8}\frac{\hbar T}{m(\Delta x_0)^2}\frac{x^2}{(\Delta x)^2}$$

The function $R(x)$ is the ground-state wave function for a simple harmonic-oscillator potential

$$U_1(x) = \frac{1}{2}\frac{h^2}{m(\Delta x)^2}\frac{x^2}{(\Delta x)^2}$$

and if Δx is to be very large, the related spring constant must be very small. The pulse potential

$$U_2(x,t) = -\hbar S(x)\,\delta(t)$$

that should then be applied is also parabolic in its space dependence.

This simple example illustrating a much more general procedure admits of a simple interpretation. The shrinking of the wave packet is accomplished when the broad Gaussian wave function $R(x)$ is acted on by a pulse potential that delivers to the electron, whatever its coördinate x might be, just the impulse required to bring it very near $x = 0$ after the passage of a time T. The procedure has certain similarities to the focusing of a beam of light by a lens.

We have now shown how to prepare an arbitrary initial state. Subsequently, this state evolves according to the time-dependent Schrödinger equation for the system of interest. If the system is isolated, its Hamiltonian is of the form $H = T + V$. We may even apply external perturbations described by known time-dependent conservative force fields, and the wave function will still develop in a causal manner. If the perturbation has some uncertain or random element, the system will no longer be describable by a wave function, but by a statistical mixture of wave functions. The density matrix of John von Neumann,[4] Lev Landau[5] and Dirac[6] provides a convenient description for such situations. However, we consider only the pure case here.

Dynamical variables

Suppose at some time t_M that we wish to make a measurement of a real dynamical variable $F(x,p)$ for the system then described by the wave function $\psi(x,t_M)$. How is this to be done in an operational manner? Is it, in fact, possible for any Hermitian operator $F(x,p)$, or just for some specially favored ones? Dirac[7] has indicated that it should be possible to measure "observables," which he defines as real dynamical variables whose eigenstates form a complete set; however Wolfgang Pauli[8] and others have doubted that the question can be decided within the framework of nonrelativistic quantum mechanics.

If $F(x,p)$ were in the form of an experimentally realizable Hamiltonian operator, one could manage to measure F very nicely as follows. Just at time t_M one would suddenly switch off $V(x)$ and instead subject the electron to the Hamiltonian $F(x,p)$. Let the eigenfunctions of $F(x,p)$ be denoted by $\phi_f(x)$ and the corresponding eigenvalues by F_f. For simplicity, we assume that any degeneracy has been lifted so that there are no repeated eigenvalues. We may then expand the wave function $\psi(x,t_M)$ in a series of the $\phi_f(x)$ as

$$\psi(x,t_M) = \sum_f C_f\,\phi_f(x)$$

At some later time $t > t_M$ the wave function would be

$$\psi(x,t) = \sum_f C_f\,\phi_f(x) \exp\left\{-(i/h)F_f(t - t_M)\right\}$$

The probability W_f of finding the system in state f, that is, having a value of the operator F equal to F_f, is given by the absolute square of the expansion coefficient C_f,

$$W_f = |C_f|^2$$

and is independent of t. One then has a long time available to decide how to determine the probability distribution over the states f. The method usually contemplated is to send the atoms through some kind of Stern–Gerlach apparatus (figure 3). This has been often discussed as in monographs and textbooks of Pauli,[9] David Bohm,[10] Kurt Gottfried[11] and others, and I do not have much to add to this part of the story. At some stage in the measurement process the wave function is converted into a mixture through the coupling of the quantum-mechanical system to a quasi-classical system, which in effect has a random interaction with the system of interest. A further stage, essentially not quantum mechanical, involves recognition that one has caught the atom, as on a photographic plate, and is required to complete the measurement process.

We have assumed here that the Hermitian operator $F(x,p)$ could be regarded as a possible Hamiltonian op-

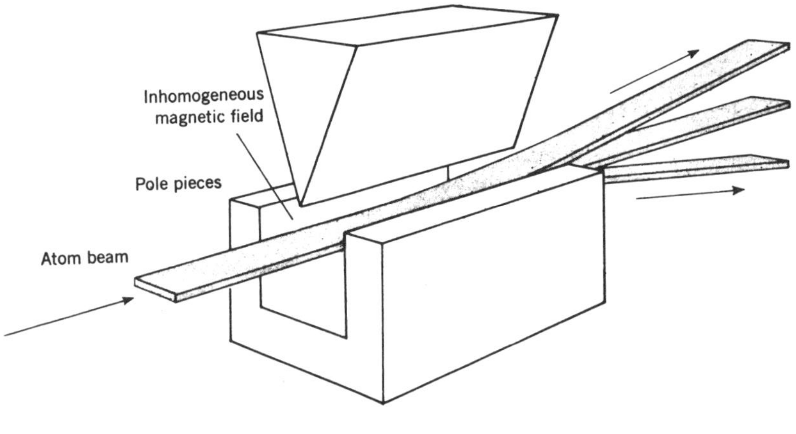

STERN–GERLACH APPARATUS. An inhomogeneous magnetic field between specially shaped pole pieces produces forces on atoms passing through; the forces are proportional to magnetic quantum number; so different states of the atoms can be separated. In some cases inhomogeneous electric fields could be used instead of these inhomogeneous magnetic fields. —FIG. 3

erator for the electron. In nonrelativistic quantum mechanics, however, in the absence of a magnetic field, all Hamiltonians are of the form $H = T + V$, where the potential V may depend on the coördinate x and time t, but not on momentum p. An easy way out is to ignore this difficulty, and to claim that the laws of quantum mechanics must prevail no matter what the Hamiltonian may be. This theoretical possibility will not suffice if we are to have an operationally possible procedure for the measurement of $F(x,p)$.

Even if $F(x,p)$ is not a Hamiltonian operator, it may still be possible to give a prescription for measuring it. We first consider some special cases of great importance. The momentum operator p is not in Hamiltonian form, but $p^2/2m$ is the Hamiltonian for a free particle. In his 1929 Chicago lectures Heisenberg[12] mentioned that the momentum of an atomic electron could be measured by suddenly turning off the atomic potential and letting the electron wave function spread. The distribution of times of arrival for electrons at a distant detector would give the desired momentum distribution.

Position coordinates

Another operator of basic importance is the coördinate x, which is also clearly not in Hamiltonian form. Nevertheless it can be measured by a simple procedure. Suppose x is to be measured at time t_M for a particle whose wave function is $\psi(x,t_M)$. That means that we want the probability distribution for finding the particle in the range x_M, $x_M + dx_M$. As usual, we have to study an ensemble of a lot of similarly prepared systems. We suddenly turn on a very strong but short-range attractive potential centered about the point x_M. Let this potential have only one bound state with energy E_1 and wave function $u_1(x - x_M)$ as shown in figure 4. The original potential $V(x)$ is simultaneously turned off. For times t later than t_M the wave function $\psi(x,t)$ can be expanded in terms of the eigenfunctions $u_n(x)$ and energy E_n, for the short-range potential

$$\psi(x,t) = \sum_n C_n u_n(x - x_M) \exp\left\{-(i/h)E_n(t - t_M)\right\}$$

where the sum has one discrete term $n = 1$ and also implies an integration over the continuous spectrum of unbound states. After a fairly short time, all parts of the wave function will spread out and become negligible near x_M except the term in $u_1(x - x_M)$. The probability amplitude C_1 is given by

$$C_1 = \int u_1(x - x_M)^* \, \psi(x,t_M) \, dx$$

and the time-independent probability for catching the electron in the potential is

$$W_1 = |C_1|^2 \qquad (4)$$

If $u_1(x - x_M)$ has a short range compared with the distance in which $\psi(x, t_M)$ varies appreciably, equation 4 becomes approximately

$$W_1 \propto |\psi(x_M,t_M)|^2$$

When this is normalized, we find that

$$dW\,(x_M) = |\psi(x_M,t_M)|^2 \, dx_M$$

in agreement with the original probability assumption of quantum mechanics. The method is similar to one that might be used to determine the distribution of flies in a room. One would quickly clasp one's fingers about a small region x_M and find out by subsequent operations whether one had caught a fly or not. Then the process would be repeated many times for similarly prepared rooms to build up a probability distribution.

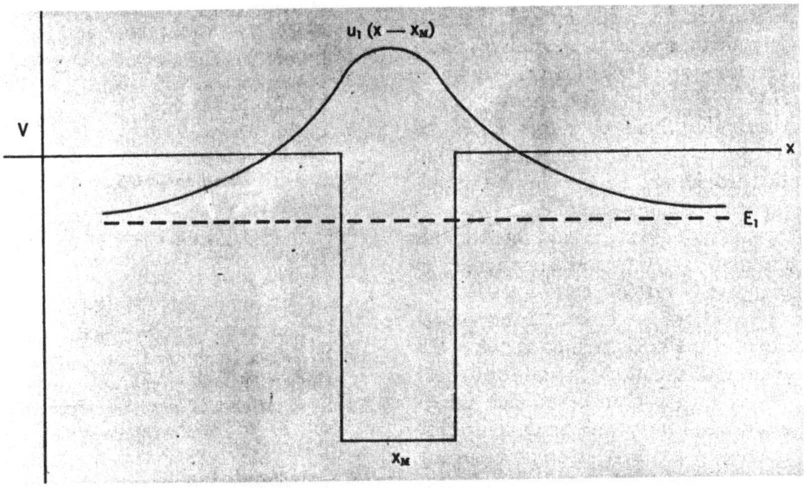

MEASUREMENT OF POSITION means determining the probability distribution for catching a particle in a short-range potential trap. This potential is centered at x_M and has one bound state $u_1(x - x_M)$ of energy E_1. The wave function describing the electron, $\psi(x,t_M)$, spreads out when the short-range potential is turned on and the original system potential $V(x)$ is turned off. All parts of the wave function quickly become negligible except for the term in $u_1(x - x_M)$, which is shown in color. —FIG. 4

Measuring Hermitian operators

In general, the non-Hamiltonian operator $F(x,p)$ will not permit itself to be measured by such simple tricks, and one might begin to doubt whether all Hermitian operators can be measured. However, the following recipe has been developed following suggestions by Yakir Aharonov of Yeshiva University. It gives a close formal, if not very physical, correspondence between the operator $F(x,p)$ and the experimental procedures to be followed. We first have to find a set of operators $G(x,p)$ that together with $F(x,p)$ form a complete set of commuting observables. This should not be hard to do in any specific case, although no general prescription is available. The simultaneous eigenfunctions of F and G should form a complete set of expansion functions spanning the Hilbert space of the problem of interest. They obey equations of the form

$$\begin{aligned} F(x,p)\ \phi_{fg}(x) &= F_f\phi_{fg}(x) \\ G(x,p)\ \phi_{fg}(x) &= G_g\phi_{fg}(x) \end{aligned} \qquad (5)$$

where the subscripts f, g label eigenfunctions and eigenvalues. (For simplicity, only one equation of type (5) is written.) The wave function $\psi(x,t_M)$ on which the measurement of F is to be made can be expanded in the form

$$\psi(x,t_M) = \sum_f \sum_g C_{fg}\phi_{fg}(x)$$

where the expansion coefficients are given by

$$C_{fg} = \int \phi_{fg}(x)^* \, \psi(x,t_M)dx\,. \qquad (6)$$

The absolute square of C_{fg} gives the probability that F has the value F_f and G has the value G_g. If we care only about F, we must carry out a sum over g.

Let us indicate an experimental procedure that can give the value of the probability distribution

$$W_{fg} = |C_{fg}|^2$$

Suppose that the eigenfunction ϕ_{fg} is written in terms of certain real functions R and S as

$$\phi_{fg}(x) = R_{fg}(x) \exp\left\{ -iS_{fg}(x) \right\}$$

where both R and S depend on the indices f and g. The first step in the measurement then involves converting the wave function $\psi(x,t_M)$ of the system into a new wave function $\psi_S(x)$ given by

$$\psi_S(x) = \psi(x,t_M) \exp\left\{ iS_{fg}(x) \right\} \quad (7)$$

by applying a pulse potential

$$U_2(x,t) = -\hbar S_{fg}(x)\, \delta(t - t_M)$$

to the system. Then the probability amplitude C_{fg} of equation 6 has the value

$$C_{fg} = \int R_{fg}(x)\, \psi_S(x)\, dx$$

which is just the overlap integral between the new wave function $\psi_S(x)$ of the system and the real normalized wave function R_{fg}. From the earlier discussion of state preparation we know how to find a potential $U_{fg}(x)$ in which $R_{fg}(x)$ is an eigenfunction of energy E. This potential is given by

$$U_{fg}(x) = E - [R_{fg}(x)]^{-1}\, T\, R_{fg}(x)$$

Hence, the second stage of the measurement process involves the sudden application of the potential $U_{fg}(x)$ and removal of the potential $V(x)$ from the system with a wave function given by equation 7. The next task is to find the probability that the particle is in the state of energy E. This could be accomplished by the usual kind of Stern–Gerlach procedure. In this manner, we find the desired probability that the operators F, G have values F_f, G_g for the state $\psi(x,t_M)$ of the system of interest. It will be noted that the measurement problem is now more complicated than for Hamiltonian operators, as the potentials $U(x,t)$ to be applied depend on the values of f and g, and hence a series of measurements has to be made for each set of f, g values.

Limitations

Some concluding comments are in order.

• We have assumed that all classically describable potentials $U(x,t)$ are available to us experimentally. This is quite similar and closely related to an assumption made by Niels Bohr and Leon Rosenfeld[13] in their discussion of the measurement of electromagnetic fields, that test bodies of very great mass and charge density were available to them, whose quantum-mechanical fluctuations of position and momentum could be neglected.

• Our discussion of measurement considers the observation of any one dynamical variable, A. Instead, we could measure another one, B, for the same state $\psi(x,t_M)$. For each operator we could determine the dispersion measure $\triangle A$ or $\triangle B$. The product of the fluctuation measures would obey the Robertson[14]–Schrödinger[15] generalization

$$\triangle A \cdot \triangle B \geq 1/2 |\langle [A,B] \rangle|$$

of Heisenberg's uncertainty relations. It should be noted, however, that such uncertainty relations do not refer to measurements, whether simultaneous or not, of a *pair* of observables. In special cases, involving commutability, it might be possible to measure first one observable and then another, but in general the thoroughgoing measurement of the first observable will so disrupt phase relations that it will serve no physical purpose to subsequently measure a second observable on the resulting mixture.

To measure simultaneously two noncommuting observables A and B (for example, x and p) one would have to find a potential $U(x,t)$ that was determined by both A and B. In general, I do not believe that this can be done in such a way that the desired information emerges from the measurement. One could, of course, form a single Hermitian operator out of the two Hermitian operators A and B. Some examples are $(AB + BA)$, $-i(AB - BA)$, $A^2B + BA^2$, ABA, etc. Any one of these Hermitian operators could be measured, as already indicated, but this would not be the desired simultaneous measurement of A and B.

• It is possible to extend the methods outlined so that measurements on many-body systems can be made.

• I do not see how to apply procedures of the kind outlined above to the relativistic quantum domain, or to field theory. In the absence of such generalizations, it may well be doubted that the story that I have given provides any significant insight into the real meaning of quantum mechanics. However, it is true that almost all expositions of quantum mechanics make use of the fictional notion that some kinds of measurements are possible. I have described certain experimental procedures for making them. There may be other ways. If they cannot be made in some fashion, either as I have suggested, or otherwise, then it appears that our understanding of the meaning of the quantum theory is correspondingly diminished, and it is only likely to be increased when a better theory of measurement for the more general relativistic and field-theoretic cases can be given. Of course, it may be that a system of rules for calculation can exist, despite the absence of an operational interpretation of the kind I have attempted. For the teaching of quantum mechanics now, it is certainly a convenient fiction to pretend that the usual textbook assumptions about measurement have a meaning, even if from an operational point of view they do not. The mathematical formulation of quantum mechanics by Dirac beautifully matches the assumed notion of measurability. However, there is clearly much more for us to learn.

* * *

This article is based on a lecture given 3 July 1968 at the 6th triennial Conference of Physics Nobel Prize Winners held at Lindau (Bodensee), West Germany. This work was supported in part by the US Air Force Office of Scientific Research.

References

1. W. Heisenberg, *Physical Principles of the Quantum Theory*, University of Chicago Press (1930), p. 21.
2. W. Pauli, "The General Principles of Wave Mechanics," in *Handbuch der Physik*, vol. 24/1, Springer, Berlin (1933), pp. 163–164.
3. E. Merzbacher, *Quantum Mechanics*, Wiley, New York (1961), p. 158.
4. J. von Neumann, Nachr. Ges. Wiss. Göttingen, p. 1 (1927); p. 245 (1927); p. 273 (1927).
5. L. D. Landau, Zeits. für Physik **45**, 430 (1927).
6. P. A. M. Dirac, Proc. Camb. Phil. Soc. **25**, 62 (1929).
7. P. A. M. Dirac, *Quantum Mechanics*, 4th ed., Oxford University Press (1958), p. 37.
8. Op. cit. ref. 2, pp. 164–166.
9. Op. cit. ref. 2, pp. 143–154.
10. D. Bohm, *Quantum Theory*, Prentice-Hall, New York (1951), chap. 22.
11. K. Gottfried, *Quantum Mechanics*, vol. 1, W. A. Benjamin, Inc., New York (1966), chap. 4.
12. Op. cit. ref. 1, p. 32.
13. N. Bohr, L. Rosenfeld, Det. Kgl. Dansk. Vid. Selskab **12**, 8 (1933).
14. H. P. Robertson, Phys. Rev. **35**, 667A (1930).
15. E. Schrödinger, Sitzungsber. preuss. Akad. Wiss., p. 296 (1930). □

Reprinted with permission from *Fundamental Questions in Quantum Mechanics,* Gordon and Breach, **5**, L.E. Ballentine, "What is the Point of the Quantum Theory of Measurement?", pp. 65-75, ©1986 London.

WHAT IS THE POINT OF THE QUANTUM THEORY OF MEASUREMENT?

L. E. Ballentine

Department of Physics, Simon Fraser University, Burnaby, B.C., Canada V5A 1S6

There are several uses of the quantum theory of measurement: (1) to criticize and clarify the interpretation of quantum mechanics (QM); (2) to examine the relation of QM to other subjects such as irreversibility; (3) to determine practial limits on precise measurements. Concerning (1) it is argued that any interpretation of QM that requires a "reduction" of the state during measurement is untenable and must be abandoned. Concerning (2) it is suggested that the theories of measurement and of irreversibility may illuminate each other through methodological analogies, but that earlier suggestions of a more direct relationship were mistaken.

I. INTRODUCTION

The literature on the quantum theory of measurement (QTM) is both voluminous and diverse, with different authors adopting different assumptions and sometimes apparently having different objectives. Wheeler and Zurek,[1] in the introduction to their recently published reprint collection, emphasize the lack of consensus about the subject. In an attempt to reduce that perceived chaos, I shall offer a preliminary classification of the literature according to its objectives, and shall emphasize some firm conclusions that have been obtained.

Although any paper that develops the possibility of a comparison between quantum mechanics (QM) and experiment could be claimed to be a part of QTM, I shall restrict QTM to the explicit analysis of the measurement process, that is to the interaction between the measurer (apparatus + observer) and the measured. Papers on QTM may be divided into three categories:

(1) Those which analyze the measurement process with the intent of criticizing or clarifying the interpretation of QM (Ex: Heisenberg's uncertainty principle (1927); Schrödinger's cat paradox (1935); numerous papers on state-vector reduction);

(2) Those which use QTM to study the interface between QM and other branches of physics (Ex: those which attempt to use QTM to explain irreversibility, or which invoke irreversibility to explain QTM);

(3) Those concerned with practical limitations on precise measurements (Ex: quantum nondemolition measurements).

Clearly one needs to understand the interpretation of QM before one can apply it reliably to other problems, and I shall have nothing to say about class (3) except that the conclusions of class (1) must be taken into account. This paper will deal primarily with (1), and very briefly with (2).

II. RIVAL INTERPRETATIONS OF QM

QTM provides a forum in which one can evaluate the relative merits of the two principal interpretations of the quantum *state* concept (or rather, the two principal classes of interpretations, since many detailed variations are possible):

(A) A pure state $|\Psi\rangle$ provides a complete and exhaustive description of an *individual* system. A dynamical variable represented by the operator Q has a value (q, say) if and only if

$$Q|\Psi\rangle = q|\Psi\rangle$$

(B) A pure (or mixed) state describes the statistical properties of an *ensemble* of similarly prepared systems.

Elsewhere[2] I have discussed these interpretations in more detail and have expounded the positive virtues of (B). Here I must confine myself to critical arguments. I shall argue that the results of QTM make (A) untenable.

III. ANALYSIS OF MEASUREMENT

Measurement involves an *object* (I) and an *apparatus* (II). We wish to measure some dynamical variable R belonging to the object. The operator corresponding to R possesses a complete set of eigenvectors,

$$R|r\rangle_{\mathrm{I}} = r|r\rangle_{\mathrm{I}}. \tag{1}$$

A general state vector for I can be expanded in terms of these eigenvectors,

$$|\psi\rangle_{\mathrm{I}} = \sum_r c_r|r\rangle_{\mathrm{I}}, \tag{2}$$

and $|c_r|^2$ is the probability that dynamical variable R takes on the value r when the state is ψ.

The apparatus (II) has a "pointer position" observable A, and a complete set of eigenvectors,

$$A|\alpha, m\rangle_{\mathrm{II}} = \alpha|\alpha, m\rangle_{\mathrm{II}}. \tag{3}$$

Here α is the "pointer position" eigenvalue, and m labels all the many other quantum numbers needed to specify an eigenvector.

The apparatus is prepared in an initial pre-measurement state, $|0, m\rangle_{\mathrm{II}}$, with $\alpha = 0$. One then introduces an interaction between I and II which must produce a unique correlation between the r value of I and the pointer position α_r of II. The properties required of the interaction are specified implicitly by placing constraints on the effect of the time development operator U.

If one requires that the measurement should not change the quantity being measured, then one must have

$$U|r\rangle_{\mathrm{I}}|0, m\rangle_{\mathrm{II}} = |r\rangle_{\mathrm{I}}|\alpha_r, m'\rangle_{\mathrm{II}}. \tag{4a}$$

Except for quantum nondemolition measurements, this is unnecessarily restrictive. In Ref. 2 I proposed

$$U|r\rangle_{\mathrm{I}}|0, m\rangle_{\mathrm{II}} = |\phi_r\rangle_{\mathrm{I}}|\alpha_r, m'\rangle_{\mathrm{II}}, \tag{4b}$$

with the final state of I being arbitrary. The purpose of this generalization was to include destructive measurements, which comprise a large fraction of real measurements. However, (4b) is still unnecessarily restrictive, in that it requires the final state to factor. The most general possibility is

$$U|r\rangle_{\rm I}|0,m\rangle_{\rm II} = \sum_{r',m'} u^{r;m}_{r',m'}|r'\rangle_{\rm I}|\alpha_r,m'\rangle_{\rm II}$$

$$= |\alpha_r;(r,m)\rangle\,, \qquad \text{say.}^3 \tag{4c}$$

The final state vector of (4c) is labelled by the "pointer position" eigenvalue α_r. The other labels (r, m) are not eigenvalues, but merely labels indicating where this vector came from via the unitary transformation U. This is the most general possibility, preserving only the one essential requirement for a successful measurement, the establishment of a unique correspondence between the initial value of r and the final pointer position α_r. The values of α_r corresponding to different r values (assumed to be discrete) should be clearly distinguishable by eye. I shall refer to these as *macroscopically distinct* values.

We now consider a general initial state (2) for the object I, which is not an eigenvector of the dynamical variable R that is being measured. Using (4c) and the linearity of the time development operator U, we obtain a final state

$$U|\psi\rangle_{\rm I}|0,m\rangle_{\rm II} = \sum_r c_r|\alpha_r,(r,m)\rangle$$

$$= |\Psi^f_m\rangle\,, \qquad \text{say.} \tag{5}$$

The final state is a *coherent superposition* of macroscopically distinct "pointer position" eigenvectors.

This fact causes no difficulties for interpretation (B), according to which the initial state describes an ensemble of similarly prepared systems[2] with $|c_r|^2$ being the probability distribution for the r values of object I, and the final state describes a correlated ensemble with $|c_r|^2$ being the probability distribution for the pointer position α_r of the apparatus II. But according to interpretation (A), it follows that since (5) is not an eigenvector of the "pointer position" operator A, the pointer has no well-defined position at the end of the measurement interaction. This is, of course, contrary to observation. So an advocate of (A) must postulate a further process, the so-called "reduction of the state vector",

$$|\Psi^f_m\rangle \rightarrow |\alpha_{r_0};(r_0,m)\rangle\,, \tag{6}$$

where r_0 is the observed value. The classic "problem of measurement" in QM is to explain this process of reduction (more accurately, projection) of the state vector.

IV. EXCUSES FOR "REDUCTION OF THE STATE VECTOR"

There have been many attempts to explain how the process of projection (6) might come about. I shall argue, not only that none of them is satisfactory, but that no such explanation is possible because the projection (6) is incompatible with the linear equation of motion. The explanations, or excuses, include the following.

4.1 The projection is caused by an unpredictable and uncontrollable disturbance of the object by the measuring apparatus.[4]

Any interaction between I and II that might cause such a disturbance is included in the interaction Hamiltonian from which the time development operator U is constructed. If the interaction satisfies the minimal condition (4c) for a successful measurement, then it must necessarily lead to the superposition (5), and not the reduced state (6).

4.2 The observer causes the reduction when he reads the result of the measurement from the apparatus.

One can avoid discussing the strangely psychic character of this proposal by simply including both apparatus and observer in the definition of II, and applying the above analysis unchanged. This method can be used to rule out any alleged external influence as a possible mechanism for state reduction.

4.3 The initial state of the apparatus II is, in practice, not fully specified as a definite pure state, as the notation $|0, m\rangle_{\mathrm{II}}$ would imply. Perhaps this imprecision in the state preparation, and the resultant uncertainty in the initial state vector, is sufficient to destroy the coherent superposition in (5).

This kind of imprecision only affects the value of m. The argument is not significantly changed if the initial state vector in (5) is replaced by an arbitrary linear combination of vectors having different m values.

4.4 In view of the imprecise specification of the initial state mentioned in 4.3, one should describe the initial state by means of a mixed state operator, and not by a pure state vector of any kind.

This is the only one of the proposed explanations which requires a non-trivial response. But before doing so in the next section, we should question its relevance. If, in accordance with interpretation (A) (which is the only reason for having considered the projection (6) in the first place), we regard a pure state vector as describing a single system (rather than an ensemble), then the use of a mixed state operator is necessitated only by our ignorance of the state vector. But as long as the state vector *exists*, even if it is unknown, the analysis of Section III applies, along with its conclusion that the final state is a coherent superposition of macroscopically distinct "pointer position" eigenvectors.

Although the "ignorance" interpretation of mixed states is commonly associated with interpretation (A), it may not be a necessary part of it, so we shall go on to an analysis of QTM with mixed states.

V. QTM FOR MIXED STATES

Instead of the initial pure state vector assumed in (5),

$$|\Psi^i_m\rangle = |\psi\rangle_{\mathrm{I}}|0, m\rangle_{\mathrm{II}},$$

we now assume an initial mixed state for the system I + II,

$$\rho^i = \sum_m w_m |\Psi^i_m\rangle\langle\Psi^i_m|. \tag{7}$$

Here w_m can be regarded as the probability associated with each of the microscopic states labelled by m (which represents many quantum numbers of the apparatus).

The hope of an advocate of interpretation (A) would now be that the final state would be a mixture of "pointer position" eigenstates, perhaps of the form

$$\rho^d = \sum_r |c_r|^2 \sum_m v_m |\alpha_r; (r, m)\rangle\langle\alpha_r; (r, m)|,$$

but certainly diagonal with respect to α_r. (Any terms that are non-diagonal in α_r would correspond to coherent superpositions of different "pointer position" eigenvectors.) A bit of reflection leads one to realize that the search for a projection of the post-measurement state of the form (6) was doomed to fail because it would have led to a unique measurement result α_{r_0}. But it is universally agreed that QM can make only probabilistic predictions, and ρ^d represents (within the confines of interpretation (A)) the prediction that the result will be α_r with probability $|c_r|^2$. Since the time development is unitary, it is impossible for an initial pure state to lead to a mixed state like ρ^d. But the use of an initial mixed state (7) appears, superficially, to offer renewed hope for interpretation (A).

That hope is very quickly dashed. The actual final state is

$$\rho^f = U\rho^i U^+ = \sum_m w_m |\Psi_m^f\rangle\langle\Psi_m^f| \tag{8}$$

where $|\Psi_m^f\rangle = U|\Psi_m^i\rangle$. Using Eq. (5), we obtain

$$\rho^f = \sum_{r_1}\sum_{r_2} c_{r_1}^* c_{r_2} \sum_m w_m |\alpha_{r_1}; (r_1, m)\rangle\langle\alpha_{r_2}; (r_2, m)|. \tag{9}$$

The terms with $\alpha_{r_1} \neq \alpha_{r_2}$ indicate *coherent superposition* of macroscopically distinct "pointer position" eigenvectors, just as was the case in (5). Since $w_m \geqslant 0$, there is no possibility that these non-diagonal (in α_r) terms can cancel out. If the minimal condition (4c) for a successful measurement is satisfied, then a coherent superposition of r-eigenvectors (2) for the object (I) must inevitably lead to a similar coherent superposition of α_r-eigenvectors for the system of object + apparatus (I + II).[5]

We have therefore proven the *conclusion*:

5.1 "Reduction" of the state of the entire system (I + II) to an incoherent mixture of "pointer position" eigenvectors is incompatible with the linear equation of motion of QM;

and its corollary,

5.2 Any interpretation of the type (A), which requires such a "reduction", is untenable.

VI. RELATION OF QTM TO IRREVERSIBILITY

There is an analogy between the QM descriptions of measurement and irreversibility, which may be schematically indicated as:

Measurement

$$\rho^i \to U\rho^i U^+ \neq \rho^d. \tag{10}$$

An initial state that is non-diagonal with respect to r (the value of the dynamical variable being measured) cannot evolve into a state that is diagonal in the apparatus "pointer position" α_r;

Irreversibility

$$\rho^{\text{non-eq}} \longrightarrow U\rho^{\text{non-eq}}U^+ \neq \rho^{\text{eq}} = e^{-\beta H}/Z. \tag{11}$$

The eigenvalues of the equilibrium state operator, ρ^{eq}, yield the Boltzmann energy distribution. Since the eigenvalues of ρ are unchanged by time development (a unitary transformation), the overwhelming majority of non-equilibrium states cannot evolve into ρ^{eq}.

It is reasonable to hope that insight into one of these problems might illuminate the other. Unfortunately it has sometimes happened that confusion about one problem has obscured the other.

Obfuscation (i): The "reduction" or "projection" of the state upon measurement would, if it occurred, be an irreversible process, since the inverse of a projection operator does not exist. Hence it is suggested that measurement or observation might be the cause of irreversibility.[6] This suggestion leads to the unacceptable conclusion that entropy would not increase (i.e. coffee would not cool, smoke would not diffuse, etc.) unless there was someone to measure it. In the light of the conclusion of Section V, we see that this suggestion and its unacceptable consequences result from the use of an unsound interpretation of QM.

Obfuscation (ii): It is asserted that equilibrium must somehow be reached by macroscopic systems. The apparatus II is macroscopic, so by invoking a transition $\rho^{\text{non-eq}} \rightarrow \rho^{\text{eq}}$ for II the non-diagonal ($\alpha_{r_1} \neq \alpha_{r_2}$) terms in the final measurement state (9) are eliminated.[7] This type of argument is the converse of (i), an imperfect understanding of irreversibility now being used to mask an inadequate theory of measurement.

A useful transfer of insight from one subject to the other may be obtained, not because the same or similar physical processes are involved in measurement and in irreversibility, but rather by means of a methodological analogy. The solution of the classic "problem of measurement" in QM, obtained by the rejection of interpretation (A) in favor of (B), did not involve any change in the mathematical formalism of QM, but rather a change in the *correspondence rules* that relate the mathematical formalism to empirical reality. Specifically, it involved a change in the interpretation of a state vector. Similarly, the conundrum of (11), which *seems* to say (but actually does not) that the attainment of equilibrium is incompatible with QM, will be resolved by a more careful analysis of the relation between the statistical state operator ρ and empirical reality. But that is another story for another day.[8]

Another methodological analogy, this time from irreversibility to QTM, is obtained from the observation that the approach to equilibrium is usually studied, not for an isolated system, but for a system in contact with a heat reservoir. If we allow the system and the reservoir to interact, and then take the trace over reservoir variables to form a partial state operator for the original system, then it may be possible to obtain

$$\mathrm{Tr}_{(\mathrm{res})} U \rho_{\mathrm{sys}}^{\mathrm{non\text{-}eq}} \otimes \rho_{\mathrm{res}} U^{+} = \rho_{\mathrm{sys}}^{\mathrm{eq}},$$

where U is the time evolution operator for the system and reservoir in interaction. This result has not been proven, but at least it is not incompatible with (11).

In our analysis of measurement we have already included all relevant external influences with the apparatus II, so there is no possibility of adding an analogue of the heat reservoir. But we may calculate partial state operators for the object I and for the apparatus II by taking partial traces of the final state operator (9) of the total system:

$$\rho_{\mathrm{I}}^{f} = \mathrm{Tr}_{(\mathrm{II})} \rho^{f}, \qquad \rho_{\mathrm{II}}^{f} = \mathrm{Tr}_{(\mathrm{I})} \rho^{f}.$$

These partial state operators may indeed be mixed states, diagonal in r and in α_r, respectively, even if the initial state is a pure state.[9] (One needs some additional properties of the coefficients $u_{r',m'}^{r,m}$ in (4c) to prove this.) But the partial state operators ρ_{I}^{f} and ρ_{II}^{f} provide only an incomplete description of the system I + II, since they do not describe the correlations between I and II. Therefore, this result is of only practical computational significance, and has no bearing on the fundamental interpretation of QM.

VII. CONCLUSIONS

The major conclusions of this paper are those stated at the end of Section V. The interaction between the object of measurement I and the apparatus II leads, in general, to a state of the system I + II that is a coherent superposition of macroscopically distinct "pointer position" eigenstates. The so-called "reduction" of the state to an eigenstate or an incoherent mixture of eigenstates is impossible within the (linear) mathematical formalism of QM. Hence any interpretation of the formalism of type (A), Section II, which regards a state vector as the complete description of an *individual* system, and so requires a "reduction", is untenable and must be discarded.

Several years[2,10] ago I advocated a *statistical ensemble* interpretation of QM that is more conservative than (A), in that it retains all the useful predictive power of QM but avoids paradoxes like the classic "problem of measurement". This interpretation also avoids the so-called QM Zeno Paradox, that a continuously observed system cannot change its state,[11] a paradox that can arise only in interpretation (A). Variations and innovations regarding the interpretation of QM are, no doubt, still possible, but whatever form they take, they must not include the fictitious process of state reduction during measurement.[12]

Notes and References

1. J. A. Wheeler and W. H. Zurek (eds.), *Quantum Theory and Measurement* (Princeton University Press, Princeton, N.J., 1983). This book is a reprint collection and large annotated bibliography.
2. L. E. Ballentine, *Rev. Mod. Phys.* **42**, 358–381 (1970).
3. Vectors without a subscript refer to the state space of the entire system I + II, whereas subscripted vectors belong to the state space of only one component, I or II.
4. This "disturbance" theory is no longer widely held. Yet a book as modern as A. Messiah, *Quantum Mechanics* (North-Holland, Amsterdam, 1964) still employs it (see p. 140).
5. This conclusion has also been reached by many other authors, by means of analyses of varying degrees of generality. Probably the most general of these is by A. Fine, *Phys. Rev.* **D2**, 2783 (1970).
6. This suggestion seems to have its origin in J. Von Neumann, *Mathematical Foundations of Quantum Mechanics* (Princeton University Press, Princeton, N.J., 1955).
7. The theory of A. Daneri, A. Loinger and G. M. Prosperi, *Nuovo Cim.* **44B**, 119 (1966) is of this type. J. Bub, *Nuovo Cim.* **57B**, 503 (1968) has given a detailed critique of their theory.
8. The necessary ideas are presented in the context of classical statistical mechanics by E. T. Jaynes, *Am. J. Phys.* **33**, 391 (1965). A translation of them into the language of quantum statistical mechanics is not difficult. It should be pointed out that Jaynes' occasional use of a subjective interpretation of probability is not essential and may be avoided.
9. H. Margenau, *Ann. Phys.* (*N.Y.*) **23**, 469 (1963).
10. Much of the credit for the ideas in Ref. 2 belongs to A. Einstein. See his "Reply to Criticisms" in *Albert Einstein: Philosopher-Scientist*, ed. P. A. Schilpp (Library of the Living Philosophers, Evanston, Ill.; and Harper-Row, New York, 1949). A

historical study of Einstein's interpretation of QM was given by L. E. Ballentine, *Am. J. Phys.* **40**, 1763 (1972).

11. J. R. Fox, "The Quantum Zeno Paradox Resolved" (preprint, 1983).
12. The first writer to stress the need to eliminate the "reduction" or "projection" postulate from QM was H. Margenau, *Phys. Rev.* **49**, 240 (1936). Some other papers that contain interpretations of QM meeting this requirement are: P. Pearle, *Am. J. Phys.* **35**, 742 (1967); R. G. Newton, *Am. J. Phys.* **48**, 1029 (1980); and J. R. Fox, *Am. J. Phys.* **51**, 49 (1983). The "Many-Worlds" interpretation meets this requirement, but fails on other grounds; see L. E. Ballentine, *Found. Phys.* **3**, 229 (1973).

Reprinted with permission from *Physics,* 1, J.S. Bell, "On the Einstein Podolsky Rosen Paradox," pp. 195-202,

Physics Vol. 1, No. 3, pp. 195–200, 1964 Physics Publishing Co. Printed in the United States

ON THE EINSTEIN PODOLSKY ROSEN PARADOX*

J. S. BELL†
Department of Physics, University of Wisconsin, Madison, Wisconsin

(*Received* 4 *November* 1964)

I. Introduction

THE paradox of Einstein, Podolsky and Rosen [1] was advanced as an argument that quantum mechanics could not be a complete theory but should be supplemented by additional variables. These additional variables were to restore to the theory causality and locality [2]. In this note that idea will be formulated mathematically and shown to be incompatible with the statistical predictions of quantum mechanics. It is the requirement of locality, or more precisely that the result of a measurement on one system be unaffected by operations on a distant system with which it has interacted in the past, that creates the essential difficulty. There have been attempts [3] to show that even without such a separability or locality requirement no "hidden variable" interpretation of quantum mechanics is possible. These attempts have been examined elsewhere [4] and found wanting. Moreover, a hidden variable interpretation of elementary quantum theory [5] has been explicitly constructed. That particular interpretation has indeed a grossly nonlocal structure. This is characteristic, according to the result to be proved here, of any such theory which reproduces exactly the quantum mechanical predictions.

II. Formulation

With the example advocated by Bohm and Aharonov [6], the EPR argument is the following. Consider a pair of spin one-half particles formed somehow in the singlet spin state and moving freely in opposite directions. Measurements can be made, say by Stern-Gerlach magnets, on selected components of the spins $\vec{\sigma}_1$ and $\vec{\sigma}_2$. If measurement of the component $\vec{\sigma}_1 \cdot \vec{a}$, where $\vec{a}$ is some unit vector, yields the value +1 then, according to quantum mechanics, measurement of $\vec{\sigma}_2 \cdot \vec{a}$ must yield the value −1 and vice versa. Now we make the hypothesis [2], and it seems one at least worth considering, that if the two measurements are made at places remote from one another the orientation of one magnet does not influence the result obtained with the other. Since we can predict in advance the result of measuring any chosen component of $\vec{\sigma}_2$, by previously measuring the same component of $\vec{\sigma}_1$, it follows that the result of any such measurement must actually be predetermined. Since the initial quantum mechanical wave function does *not* determine the result of an individual measurement, this predetermination implies the possibility of a more complete specification of the state.

Let this more complete specification be effected by means of parameters λ. It is a matter of indifference in the following whether λ denotes a single variable or a set, or even a set of functions, and whether the variables are discrete or continuous. However, we write as if λ were a single continuous parameter. The result A of measuring $\vec{\sigma}_1 \cdot \vec{a}$ is then determined by $\vec{a}$ and λ, and the result B of measuring $\vec{\sigma}_2 \cdot \vec{b}$ in the same instance is determined by $\vec{b}$ and λ, and

*Work supported in part by the U.S. Atomic Energy Commission
†On leave of absence from SLAC and CERN

$$A(\vec{a}, \lambda) = \pm 1, \; B(\vec{b}, \lambda) = \pm 1. \tag{1}$$

The vital assumption [2] is that the result B for particle 2 does not depend on the setting $\vec{a}$, of the magnet for particle 1, nor A on $\vec{b}$.

If $\rho(\lambda)$ is the probability distribution of λ then the expectation value of the product of the two components $\vec{\sigma}_1 \cdot \vec{a}$ and $\vec{\sigma}_2 \cdot \vec{b}$ is

$$P(\vec{a}, \vec{b}) = \int d\lambda \, \rho(\lambda) A(\vec{a}, \lambda) B(\vec{b}, \lambda) \tag{2}$$

This should equal the quantum mechanical expectation value, which for the singlet state is

$$\langle \vec{\sigma}_1 \cdot \vec{a} \; \vec{\sigma}_2 \cdot \vec{b} \rangle = -\vec{a} \cdot \vec{b}\,. \tag{3}$$

But it will be shown that this is not possible.

Some might prefer a formulation in which the hidden variables fall into two sets, with A dependent on one and B on the other; this possibility is contained in the above, since λ stands for any number of variables and the dependences thereon of A and B are unrestricted. In a complete physical theory of the type envisaged by Einstein, the hidden variables would have dynamical significance and laws of motion; our λ can then be thought of as initial values of these variables at some suitable instant.

III. Illustration

The proof of the main result is quite simple. Before giving it, however, a number of illustrations may serve to put it in perspective.

Firstly, there is no difficulty in giving a hidden variable account of spin measurements on a single particle. Suppose we have a spin half particle in a pure spin state with polarization denoted by a unit vector $\vec{p}$. Let the hidden variable be (for example) a unit vector $\vec{\lambda}$ with uniform probability distribution over the hemisphere $\vec{\lambda} \cdot \vec{p} > 0$. Specify that the result of measurement of a component $\vec{\sigma} \cdot \vec{a}$ is

$$\text{sign } \vec{\lambda} \cdot \vec{a}'\,, \tag{4}$$

where $\vec{a}'$ is a unit vector depending on $\vec{a}$ and $\vec{p}$ in a way to be specified, and the sign function is $+1$ or -1 according to the sign of its argument. Actually this leaves the result undetermined when $\lambda \cdot a' = 0$, but as the probability of this is zero we will not make special prescriptions for it. Averaging over $\vec{\lambda}$ the expectation value is

$$\langle \vec{\sigma} \cdot \vec{a} \rangle = 1 - 2\theta'/\pi\,, \tag{5}$$

where θ' is the angle between $\vec{a}'$ and $\vec{p}$. Suppose then that $\vec{a}'$ is obtained from $\vec{a}$ by rotation towards $\vec{p}$ until

$$1 - \frac{2\theta'}{\pi} = \cos\theta \tag{6}$$

where θ is the angle between $\vec{a}$ and $\vec{p}$. Then we have the desired result

$$\langle \vec{\sigma} \cdot \vec{a} \rangle = \cos\theta \tag{7}$$

So in this simple case there is no difficulty in the view that the result of every measurement is determined by the value of an extra variable, and that the statistical features of quantum mechanics arise because the value of this variable is unknown in individual instances.

Secondly, there is no difficulty in reproducing, in the form (2), the only features of (3) commonly used in verbal discussions of this problem:

$$\left.\begin{aligned} P(\vec{a}, \vec{a}) &= -P(\vec{a}, -\vec{a}) = -1 \\ P(\vec{a}, \vec{b}) &= 0 \text{ if } \vec{a}\cdot\vec{b} = 0 \end{aligned}\right\} \tag{8}$$

For example, let λ now be unit vector $\vec{\lambda}$, with uniform probability distribution over all directions, and take

$$\left.\begin{aligned} A(\vec{a}, \vec{\lambda}) &= \text{sign } \vec{a}\cdot\vec{\lambda} \\ B(a, b) &= -\text{sign } \vec{b}\cdot\vec{\lambda} \end{aligned}\right\} \tag{9}$$

This gives

$$P(\vec{a}, \vec{b}) = -1 + \frac{2}{\pi}\theta, \tag{10}$$

where θ is the angle between a and b, and (10) has the properties (8). For comparison, consider the result of a modified theory [6] in which the pure singlet state is replaced in the course of time by an isotropic mixture of product states; this gives the correlation function

$$-\frac{1}{3}\vec{a}\cdot\vec{b} \tag{11}$$

It is probably less easy, experimentally, to distinguish (10) from (3), than (11) from (3).

Unlike (3), the function (10) is not stationary at the minimum value -1 (at $\theta = 0$). It will be seen that this is characteristic of functions of type (2).

Thirdly, and finally, there is no difficulty in reproducing the quantum mechanical correlation (3) if the results A and B in (2) are allowed to depend on $\vec{b}$ and $\vec{a}$ respectively as well as on $\vec{a}$ and $\vec{b}$. For example, replace $\vec{a}$ in (9) by $\vec{a}'$, obtained from $\vec{a}$ by rotation towards $\vec{b}$ until

$$1 - \frac{2}{\pi}\theta' = \cos\theta,$$

where θ' is the angle between $\vec{a}'$ and $\vec{b}$. However, for given values of the hidden variables, the results of measurements with one magnet now depend on the setting of the distant magnet, which is just what we would wish to avoid.

IV. Contradiction

The main result will now be proved. Because ρ is a normalized probability distribution,

$$\int d\lambda\, \rho(\lambda) = 1, \tag{12}$$

and because of the properties (1), P in (2) cannot be less than -1. It can reach -1 at $\vec{a} = \vec{b}$ only if

$$A(\vec{a}, \lambda) = -B(\vec{a}, \lambda) \tag{13}$$

except at a set of points λ of zero probability. Assuming this, (2) can be rewritten

$$P(\vec{a}, \vec{b}) = -\int d\lambda\, \rho(\lambda)\, A(\vec{a}, \lambda)\, A(\vec{b}, \lambda). \tag{14}$$

It follows that if $\vec{c}$ is another unit vector then

$$P(\vec{a}, \vec{b}) - P(\vec{a}, \vec{c}) = -\int d\lambda \rho(\lambda) [A(\vec{a}, \lambda) A(\vec{b}, \lambda) - A(\vec{a}, \lambda) A(\vec{c}, \lambda)]$$
$$= \int d\lambda \rho(\lambda) A(\vec{a}, \lambda) A(\vec{b}, \lambda) [A(\vec{b}, \lambda) A(\vec{c}, \lambda) - 1]$$

using (1), whence

$$|P(\vec{a}, \vec{b}) - P(\vec{a}, \vec{c})| \leq \int d\lambda \rho(\lambda) [1 - A(\vec{b}, \lambda) A(\vec{c}, \lambda)]$$

The second term on the right is $P(\vec{b}, \vec{c})$, whence

$$1 + P(\vec{b}, \vec{c}) \geq |P(\vec{a}, \vec{b}) - P(\vec{a}, \vec{c})| \tag{15}$$

Unless P is constant, the right hand side is in general of order $|\vec{b}-\vec{c}|$ for small $|\vec{b}-\vec{c}|$. Thus $P(\vec{b}, \vec{c})$ cannot be stationary at the minimum value (-1 at $\vec{b} = \vec{c}$) and cannot equal the quantum mechanical value (3).

Nor can the quantum mechanical correlation (3) be arbitrarily closely approximated by the form (2). The formal proof of this may be set out as follows. We would not worry about failure of the approximation at isolated points, so let us consider instead of (2) and (3) the functions

$$\bar{P}(\vec{a}, \vec{b}) \quad \text{and} \quad \overline{-\vec{a} \cdot \vec{b}}$$

where the bar denotes independent averaging of $P(\vec{a}', \vec{b}')$ and $-\vec{a}' \cdot \vec{b}'$ over vectors $\vec{a}'$ and $\vec{b}'$ within specified small angles of $\vec{a}$ and $\vec{b}$. Suppose that for all $\vec{a}$ and $\vec{b}$ the difference is bounded by ϵ:

$$|\bar{P}(\vec{a}, \vec{b}) + \overline{\vec{a} \cdot \vec{b}}| \leq \epsilon \tag{16}$$

Then it will be shown that ϵ cannot be made arbitrarily small.

Suppose that for all a and b

$$|\overline{\vec{a} \cdot \vec{b}} - \vec{a} \cdot \vec{b}| \leq \delta \tag{17}$$

Then from (16)

$$|\bar{P}(\vec{a}, \vec{b}) + \vec{a} \cdot \vec{b}| \leq \epsilon + \delta \tag{18}$$

From (2)

$$\bar{P}(\vec{a}, \vec{b}) = \int d\lambda \rho(\lambda) \bar{A}(\vec{a}, \lambda) \bar{B}(\vec{b}, \lambda) \tag{19}$$

where

$$|\bar{A}(\vec{a}, \lambda)| \leq 1 \quad \text{and} \quad |\bar{B}(\vec{b}, \lambda)| \leq 1 \tag{20}$$

From (18) and (19), with $\vec{a} = \vec{b}$,

$$\int d\lambda \rho(\lambda) [\bar{A}(\vec{b}, \lambda) \bar{B}(\vec{b}, \lambda) + 1] \leq \epsilon + \delta \tag{21}$$

From (19)

$$\bar{P}(\vec{a}, \vec{b}) - \bar{P}(\vec{a}, \vec{c}) = \int d\lambda \rho(\lambda) [\bar{A}(\vec{a}, \lambda) \bar{B}(\vec{b}, \lambda) - \bar{A}(\vec{a}, \lambda) \bar{B}(\vec{c}, \lambda)]$$
$$= \int d\lambda \rho(\lambda) \bar{A}(\vec{a}, \lambda) \bar{B}(\vec{b}, \lambda) [1 + \bar{A}(\vec{b}, \lambda) \bar{B}(\vec{c}, \lambda)]$$
$$- \int d\lambda \rho(\lambda) \bar{A}(\vec{a}, \lambda) \bar{B}(\vec{c}, \lambda) [1 + \bar{A}(\vec{b}, \lambda) \bar{B}(\vec{b}, \lambda)]$$

Using (20) then

$$|\bar{P}(\vec{a}, \vec{b}) - \bar{P}(\vec{a}, \vec{c})| \leq \int d\lambda \rho(\lambda)\,[1 + \bar{A}(\vec{b}, \lambda)\,\bar{B}(\vec{c}, \lambda)] + \int d\lambda \rho(\lambda)\,[1 + \bar{A}(\vec{b}, \lambda)\,\bar{B}(\vec{b}, \lambda)]$$

Then using (19) and 21)

$$|\bar{P}(\vec{a}, \vec{b}) - \bar{P}(\vec{a}, \vec{c})| \leq 1 + \bar{P}(\vec{b}, \vec{c}) + \epsilon + \delta$$

Finally, using (18),

$$|\vec{a} \cdot \vec{c} - \vec{a} \cdot \vec{b}| - 2(\epsilon + \delta) \leq 1 - \vec{b} \cdot \vec{c} + 2(\epsilon + \delta)$$

or

$$4(\epsilon + \delta) \geq |\vec{a} \cdot \vec{c} - \vec{a} \cdot \vec{b}| + \vec{b} \cdot \vec{c} - 1 \qquad (22)$$

Take for example $\vec{a} \cdot \vec{c} = 0$, $\vec{a} \cdot \vec{b} = \vec{b} \cdot \vec{c} = 1/\sqrt{2}$ Then

$$4(\epsilon + \delta) \geq \sqrt{2} - 1$$

Therefore, for small finite δ, ϵ cannot be arbitrarily small.

Thus, the quantum mechanical expectation value cannot be represented, either accurately or arbitrarily closely, in the form (2).

V. Generalization

The example considered above has the advantage that it requires little imagination to envisage the measurements involved actually being made. In a more formal way, assuming [7] that any Hermitian operator with a complete set of eigenstates is an "observable", the result is easily extended to other systems. If the two systems have state spaces of dimensionality greater than 2 we can always consider two dimensional subspaces and define, in their direct product, operators $\vec{\sigma}_1$ and $\vec{\sigma}_2$ formally analogous to those used above and which are zero for states outside the product subspace. Then for at least one quantum mechanical state, the "singlet" state in the combined subspaces, the statistical predictions of quantum mechanics are incompatible with separable predetermination.

VI. Conclusion

In a theory in which parameters are added to quantum mechanics to determine the results of individual measurements, without changing the statistical predictions, there must be a mechanism whereby the setting of one measuring device can influence the reading of another instrument, however remote. Moreover, the signal involved must propagate instantaneously, so that such a theory could not be Lorentz invariant.

Of course, the situation is different if the quantum mechanical predictions are of limited validity. Conceivably they might apply only to experiments in which the settings of the instruments are made sufficiently in advance to allow them to reach some mutual rapport by exchange of signals with velocity less than or equal to that of light. In that connection, experiments of the type proposed by Bohm and Aharonov [6], in which the settings are changed during the flight of the particles, are crucial.

I am indebted to Drs. M. Bander and J. K. Perring for very useful discussions of this problem. The first draft of the paper was written during a stay at Brandeis University; I am indebted to colleagues there and at the University of Wisconsin for their interest and hospitality.

References

1. A. EINSTEIN, N. ROSEN and B. PODOLSKY, *Phys. Rev.* **47**, 777 (1935); see also N. BOHR, *Ibid.* **48**, 696 (1935), W. H. FURRY, *Ibid.* **49**, 393 and 476 (1936), and D. R. INGLIS, *Rev. Mod. Phys.* **33**, 1 (1961).
2. "But on one supposition we should, in my opinion, absolutely hold fast: the real factual situation of the system S_2 is independent of what is done with the system S_1, which is spatially separated from the former." A. EINSTEIN in *Albert Einstein, Philosopher Scientist,* (Edited by P. A. SCHILP) p. 85, Library of Living Philosophers, Evanston, Illinois (1949).
3. J. VON NEUMANN, *Mathematishe Grundlagen der Quanten-mechanik.* Verlag Julius-Springer, Berlin (1932), [English translation: Princeton University Press (1955)]; J. M. JAUCH and C. PIRON, *Helv. Phys. Acta* **36**, 827 (1963).
4. J. S. BELL, to be published.
5. D. BOHM, *Phys. Rev.* **85**, 166 and 180 (1952).
6. D. BOHM and Y. AHARONOV, *Phys. Rev.* **108**, 1070 (1957).
7. P. A. M. DIRAC, *The Principles of Quantum Mechanics* (3rd Ed.) p. 37. The Clarendon Press, Oxford (1947).

Physics Vol. 1, No. 3, pp. 201-202, 1964. Physics Publishing Co. Printed in the United States. 201

IGNORE

ERRATA

"The Magnetic Properties of Superconducting Alloys I," by Kazumi Maki, Vol. 1, No. 1, pp. 21–30.

Eq. (13) $K_1(T) = K(1 + 0.391\,\theta)$

should be replaced by

$$= K(1 + 0.13\,\theta),$$

and Eq. (23) $K_2 = K(1 - 0.087\,\theta)$

should be replaced by

$$= K(1 - 0.85\,\theta)\,.$$

The author would like to thank Dr. B. B. Goodman for pointing out a numerical error.

"The Magnetic Properties of Superconducting Alloys II," by Kazumi Maki, Vol. 1, No. 2, pp. 127–143.

Eq. (45) should be corrected as follows

$$F = -\frac{1}{4\pi}\cdot\frac{H_{c2}(T)(H_{c2}(T) - H)}{(2\kappa_2^2(T) - 1)}, \qquad F = -\frac{1}{8\pi}\left(H_0^2 + \frac{(H_{c2}(T) - H_0)^2}{(2\kappa_2^2(T) - 1)\beta}\right)$$

and consequently Eq. (46) should be corrected as follows

$$C_s(T) = -T\frac{\partial^2 F}{\partial T^2} \cong \frac{T}{4\pi}\frac{1}{(2\kappa_2^2(T) - 1)\beta}\left\{\left(\frac{\partial H_{c2}}{\partial T}\right)^2 + \frac{\partial^2 H_{c2}}{\partial T^2}(H_{c2} - H_0)\right\}$$

$$= \frac{16}{9}\,mp_0\frac{\kappa_1^2(0)}{(2\kappa_2^2(T) - 1)\beta}\left\{\left(\frac{\pi T}{\Delta_\infty}\right)^2 - \frac{3}{4}\frac{H_{c2}(T) - H_0}{H_{c2}(0)}\right\}T, \qquad \text{for } T \ll T_{co}.$$

IGNORE

Also, Eqs. (73), (74) and (75) should read as follows:

$$C_s(T) = T\frac{\partial^2}{\partial T^2}\left\{\frac{1}{8\pi}\frac{(H_{c2}(T) - H_0)^2}{(2\kappa_2^{*2}(T) - 1)\beta}\right\},$$

$$\cong \frac{T}{4\pi}\frac{\left(\dfrac{\partial H_{c2}}{\partial T}\right)^2}{(2\kappa_2^{*2}(T) - 1)\beta},$$

$$= \frac{mp_0}{9}\frac{16\kappa_1^{*2}(0)}{(2\kappa_2^{*2}(T) - 1)\beta}\left(\frac{1 - a^2}{1 + a^2}\right)^2\left(\frac{\pi T}{\Delta_\infty}\right)^2 T, \qquad \text{for } T \ll T_{co}$$

$$= \frac{8mp_0}{7\zeta(3)}\left\{\frac{\kappa^2}{(2\kappa^2 - 1)\beta}\right\}T, \qquad \text{for } T_{co} - T \ll T_{co}$$

respectively.

Experimental Test of Local Hidden-Variable Theories*

Stuart J. Freedman and John F. Clauser
Department of Physics and Lawrence Berkeley Laboratory, University of California, Berkeley, California 94720
(Received 4 February 1972)

We have measured the linear polarization correlation of the photons emitted in an atomic cascade of calcium. It has been shown by a generalization of Bell's inequality that the existence of local hidden variables imposes restrictions on this correlation in conflict with the predictions of quantum mechanics. Our data, in agreement with quantum mechanics, violate these restrictions to high statistical accuracy, thus providing strong evidence against local hidden-variable theories.

Since quantum mechanics was first developed, there have been repeated suggestions that its statistical features possibly might be described by an underlying deterministic substructure. Such features, then, arise because a quantum state represents a statistical ensemble of "hidden-variable states." Proofs by von Neumann and others, demonstrating the impossibility of a hid-

VOLUME 28, NUMBER 14 PHYSICAL REVIEW LETTERS 3 APRIL 1972

den-variable substructure consistent with quantum mechanics, rely on various assumptions concerning the character of the hidden variables.[1] Bell has argued that these assumptions are unduly restrictive. However, by considering an idealized case of two spatially separated but quantum-mechanically correlated systems, he was able to show that any hidden-variable theory satisfying only the natural assumption of "locality" also leads to predictions ("Bell's inequality") in conflict with quantum mechanics.[2]

Bell's proof was extended to realizable systems by Clauser, Horne, Shimony, and Holt,[3] who also pointed out that their generalization of Bell's inequality can be tested experimentally, thus testing all local hidden-variable theories, but that existing experimental results were insufficient for this purpose. This Letter reports the results of an experiment which are sufficiently precise to rule out local hidden-variable theories with high statistical accuracy

In the present work we measured the correlation in linear polarization of two photons γ_1 and γ_2 emitted in a $J=0 \rightarrow J=1 \rightarrow J=0$ atomic cascade. The decaying atoms were viewed by two symmetrically placed optical systems, each consisting of two lenses, a wavelength filter, a rotatable and removable polarizer, and a single-photon detector (see Fig. 1). The following quantities were measured: $R(\varphi)$, the coincidence rate for two-photon detection, as a function of the angle φ between the planes of linear polarization defined by the orientation of the inserted polarizers; R_1, the coincidence rate with polarizer 2 removed; R_2, the coincidence rate with polarizer 1 removed[4]; R_0, the coincidence rate with both polarizers removed. Quantum mechanics predicts that $R(\varphi)$ and R_0 are related as follows[3,5]:

$$R(\varphi)/R_0 = \tfrac{1}{4}(\epsilon_M{}^1 + \epsilon_m{}^1)(\epsilon_M{}^2 + \epsilon_m{}^2) + \tfrac{1}{4}(\epsilon_M{}^1 - \epsilon_m{}^1) \times (\epsilon_M{}^2 - \epsilon_m{}^2) F_1(\theta)\cos 2\varphi, \qquad (1a)$$

while

$$R_1/R_0 = \tfrac{1}{2}(\epsilon_M{}^1 + \epsilon_m{}^1), \qquad (1b)$$

and

$$R_2/R_0 = \tfrac{1}{2}(\epsilon_M{}^2 + \epsilon_m{}^2). \qquad (1c)$$

Here $\epsilon_M{}^i$ ($\epsilon_m{}^i$) is the transmittance of the ith polarizer for light polarized parallel (perpendicular) to the polarizer axis, and $F_1(\theta)$ is a function of the half-angle θ subtended by the primary lenses. It represents a depolarization due to noncollinearity of the two photons, and approaches unity for infinitesimal detector solid angles. [For this experiment, $\theta \approx 30°$, and $F_1(30°) \approx 0.99$.]

We make the following assumptions for any local hidden-variable theory: (1) The two photons propagate as separated localized particles. (2) A binary selection process occurs for each photon at each polarizer (transmission or no-transmission). This selection does not depend upon the orientation of the distant polarizer.

In addition, we make the following assumption to allow a comparison of the generalization of Bell's inequality with out experiment: (3) All photons incident on a detector have a probability of detection that is independent of whether or not the photon has passed through a polarizer.[6]

The above assumptions constrain the coincidence rates by the following inequalities[7]:

$$-1 \le \Delta(\varphi) \le 0, \qquad (2)$$

where

$$\Delta(\varphi) = \frac{3R(\varphi)}{R_0} - \frac{R(3\varphi)}{R_0} - \frac{R_1 + R_2}{R_0}.$$

For sufficiently small detector solid angles and highly efficient polarizers, these inequalities (2) are not satisfied by the quantum-mechanical prediction (1) for a range of values of φ. Maximum violations occur at $\varphi = 22\tfrac{1}{2}°$ $[\Delta(\varphi) > 0]$ and $\varphi = 67\tfrac{1}{2}°$ $[\Delta(\varphi) < -1]$. At these angles of maximum violation, inequalities (2) can be combined into the simpler and more convenient expression

$$\delta = |R(22\tfrac{1}{2}°)/R_0 - R(67\tfrac{1}{2}°)/R_0| - \tfrac{1}{4} \le 0, \qquad (3)$$

which does not involve R_1 or R_2.

The experimental arrangement was similar to

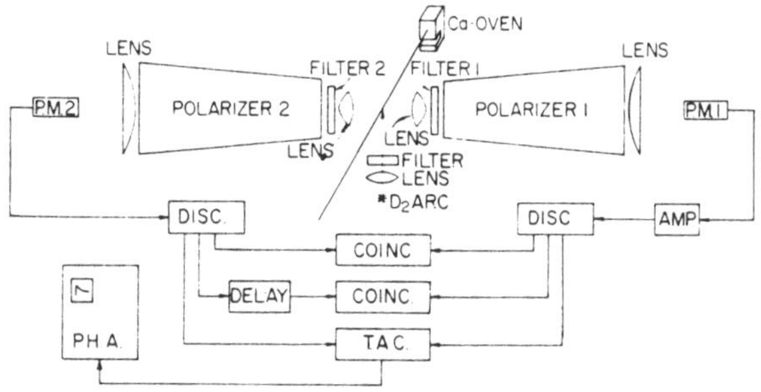

FIG. 1. Schematic diagram of apparatus and associated electronics. Scalers (not shown) monitored the outputs of the discriminators and coincidence circuits during each 100-sec count period. The contents of the scalers and the experimental configuration were recorded on paper tape and analyzed on an IBM 1620-II computer.

VOLUME 28, NUMBER 14 PHYSICAL REVIEW LETTERS 3 APRIL 1972

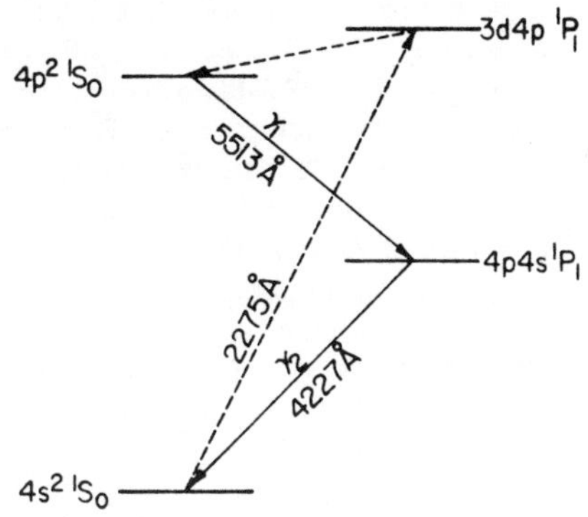

FIG. 2. Level scheme of calcium. Dashed lines show the route for excitation to the initial state $4p^2\ ^1S_0$.

that of Kocher and Commins.[8] A calcium atomic beam effused from a tantalum oven, as shown in Fig. 1. The continuum output of a deuterium arc lamp (ORIEL C-42-72-12) was passed through an interference filter [250 Å full width at half-maximum (FWHM), 20% transmission at 2275 Å] and focused on the beam. Resonance absorption of a 2275-Å photon excited calcium atoms to the $3d4p\ ^1P_1$ state. Of the atoms that did not decay directly to the ground state, about 7% decayed to the $4p^2\ ^1S_0$ state, from which they cascaded through the $4s4p\ ^1P_1$ intermediate state to the ground state with the emission of two photons at 5513 Å (γ_1) and 4227 Å (γ_2) (see Fig. 2). At the interaction region (roughly, a cylinder 5 mm high and 3 mm in diameter) the density of the calcium was about 1×10^{10} atoms/cm^3. To avoid spherical aberrations which would have reduced counter efficiencies, aspheric primary lenses (8.0 cm diam, $f=0.8$) were used. Photons γ_1 were selected by a filter with 10 Å FWHM and 50% transmission, and γ_2 by a filter with 6 Å FWHM and 20% transmission. The requirement for large efficient linear polarizers led us to employ "pile-of-plates" polarizers. Each polarizer consisted of ten 0.3-mm-thick glass sheets inclined nearly at Brewster's angle. The sheets were attached to hinged frames, and could be folded completely out of the optical path. A Geneva mechanism rotated each polarizer through increments of $22\frac{1}{2}°$. The measured transmittances of the polarizers were $\epsilon_M{}^1=0.97\pm0.01$, $\epsilon_m{}^1=0.038\pm0.004$, $\epsilon_M{}^2=0.96\pm0.01$, and $\epsilon_m{}^2=0.037\pm0.004$. The photomultiplier detectors (RCA C31000E, quantum efficiency ≈ 0.13 at 5513 Å; and RCA 8850, quantum efficiency ≈ 0.28 at 4227 Å) were cooled, reducing dark rates to 75 and 200 counts/sec, respectively. The measured counter efficiencies with polarizers removed were $\eta_1\approx1.7\times10^{-3}$ and $\eta_2\approx1.5\times10^{-3}$.[9]

A diagram of the electronics is included in Fig. 1. The overall system time resolution was about 1.5 nsec. The short intermediate state lifetime (~5 nsec) permitted a narrow coincidence window (8.1 nsec). A second coincidence channel displaced in time by 50 nsec monitored the number of accidental coincidences, the true coincidence rate being determined by subtraction.[10] A time-to-amplitude converter and pulse-height analyzer measured the time-delay spectrum of the two photons. The resulting exponential gave the intermediate state lifetime.[11]

The coincidence rates depended upon the beam and lamp intensities, the latter gradually decreasing during a run. The typical coincidence rate with polarizers removed ranged from 0.3 to 0.1 countx/sec, and the accidental rate ranged from 0.01 to 0.002 counts/sec. Long runs required by the low coincidence rate necessitated automatic data collections.

The system was cycled with 100-sec counting periods. Periods with one or both polarizers inserted alternated with periods in which both polarizers were removed. Both polarizers rotated according to a prescribed sequence. For a given run, $R(\varphi)/R_0$ was calculated by summing counts for all configurations corresponding to angle φ and dividing by half the sum of the counts in the adjacent periods of the sequence in which both polarizers were moved. Data for R_1/R_0 and R_2/R_0 were analyzed in a similar fashion. The values given here are averages over the orientation of the inserted polarizer. This cycling and averaging procedure minimized the effects of drift and apparatus asymmetry.

The results of the measurements of the correlation $R(\varphi)/R_0$, corresponding to a total integration time of ~200 h, are shown in Fig. 3. All error limits are conservative estimates of 1 standard deviation. Using the values at $22\frac{1}{2}°$ and $67\frac{1}{2}°$, we obtain $\delta=0.050\pm0.008$ in clear violation of inequality (3).[12] Furthermore, we observe no evidence for a deviation from the predictions of quantum mechanics, calculated from the measured polarizer efficiences and solid angles, and shown as the solid curve in Fig. 3. We consider these results to be strong evidence against local hidden-variable theories.

The authors wish to express their sincerest appreciation for guidance and help from Professor Eugene Commins, to Professor Charles Townes for his encouragement of this work, and to M. Sim-

VOLUME 28, NUMBER 14 PHYSICAL REVIEW LETTERS 3 APRIL 1972

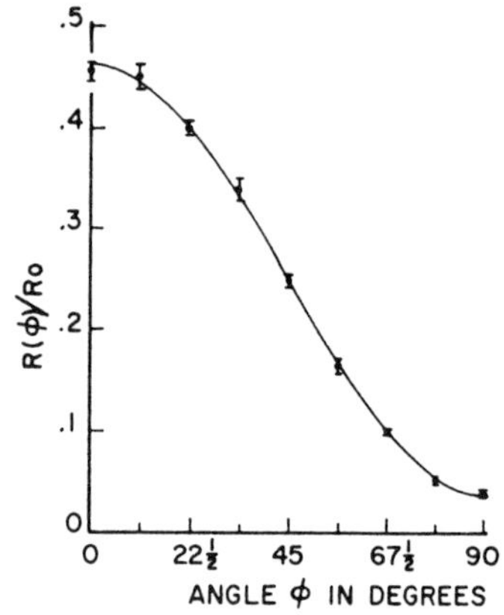

FIG. 3. Coincidence rate with angle φ between the polarizers, divided by the rate with both polarizers removed, plotted versus the angle φ. The solid line is the prediction by quantum mechanics, calculated using the measured efficiencies of the polarizers and solid angles of the experiment.

mons for helpful suggestions.

---

*Work supported by U. S. Atomic Energy Commission.

[1]The best-known proof is by J. von Neumann, *Mathematische Grundlagen der Quantemechanik* (Springer, Berlin, 1932) [*Mathematical Foundations of Quantum Mechanics* (Princeton Univ. Press. Princeton, N. J., 1955)]. For a critical review of this and other proofs see J. S. Bell, Rev. Mod. Phys. 38, 447 (1966).

[2]J. S. Bell, Physics (Long Is. City, N.Y.) 1, 195 (1964).

[3]J. Clauser, M. Horne, A. Shimony, and R. Holt, Phys. Rev. Lett. 29, 000 (1960).

[4]A hidden-variable theory need not require that R_1 and R_2 be independent of the orientation of the inserted polarizer, and we do not assume this independence in our data analysis. However, the results are consistent with R_1 and R_2 being independent of angle, and for simplicity they are so denoted.

[5]M. Horne, Ph. D. thesis, Boston University, 1970 (unpublished). See also A. Shimony, in "Foundations of Quantum Mechanics, Proceedings of the International School of Physics 'Enrico Fermi,' Course IL" (Academic, New York, to be published).

[6]This assumption is much weaker than the assumption made by L. R. Kasday, J. Ullman, and C. S. Wu, Bull. Amer. Phys. Soc. 15, 586 (1970), in their discussion of the two-γ decay of positronium; see L. R. Kasday, in "Foundations of Quantum Mechanics, Proceedings of the International School of Physics 'Enrico Fermi,' Course IL" (Academic, New York, to be published).

[7]The inequality $\Delta(\varphi) \leq 0$ is derived in Refs. 3 and 5. The other forms of the hidden-variable restriction are obtained by similar arguments; see S. Freedman, Ph. D. thesis, University of California, Berkeley, Lawrence Berkeley Laboratory Report No. LBL-391, 1972 (unpublished).

[8]C. A. Kocher and E. D. Commins, Phys. Rev. Lett. 18, 575 (1967); C. A. Kocher, Ph. D. thesis, University of California, Berkeley, Lawrence Berkeley Laboratory Report No. UCRL-17587, 1967 (unpublished).

[9]The counter efficiencies are given by $\eta_i = (\Omega_i/4\pi)T_i \times \epsilon_i L_i$, where Ω_i is the solid angle, T_i is the transmission of the filter, ϵ_i is the quantum efficiency, and L_i accounts for other losses. The measurement of η_2 was made, employing the properties of the calcium cascade, by comparing the coincidence rate and the γ_1 singles rate after suitable background correction; η_1 was then inferred from the known quantum efficiencies and filter transmissions assuming that Ω_i and L_i were the same for both detector systems.

[10]An estimate of the accidental rate was also obtained from the singles rates. The two estimates gave consistent results. In fact, our conclusions are not changed if accidentals are neglected entirely; the signal-to-accidental ratio with polarizer removed is about 40 to 1 for the data presented.

[11]Resonance trapping, encountered at high beam densities, resulted in a lengthening of the observed lifetime and a slight decrease in the polarization correlation amplitude, see J. P. Barrat, J. Phys. Radium 20, 541, 633 (1959). At low beam densities the measured lifetime is consistent with previously measured values. See W. L. Weise, M. W. Smith, and B. M. Miles, *Atomic Transition Probabilities*, U. S. National Bureau of Standards Reference Data Series—22 (U.S. GPO, Washington, D.C., 1969), Vol. 2.

[12]The results that are of interest in comparison with the hidden-variable inequalities are $R_1/R_0 = 0.497 \pm 0.009$, $R_2/R_0 = 0.499 \pm 0.009$, $R(22\frac{1}{2}°)/R_0 = 0.400 \pm 0.007$, and $R(67\frac{1}{2}°)/R_0 = 0.100 \pm 0.003$. We thus obtain $\Delta(22\frac{1}{2}°) = 0.104 \pm 0.026$ and $\Delta(67\frac{1}{2}°) = -1.097 \pm 0.018$, in violation of inequalities (2).

VOLUME 28, NUMBER 14 PHYSICAL REVIEW LETTERS 3 APRIL 1972

IGNORE

ERRATA

ENERGY SPECTRUM OF AN ELECTRON IN A PERIODIC DEFORMABLE LATTICE. David Emin [Phys. Rev. Lett. 28, 604 (1972)].

Equation (4) should read $(E_b/6J)(a/R_p)^2 < K$, and the first complete sentence below Eq. (4) should be replaced by the following: "Condition (4) insures that the weak-coupling solution is energetically stable with respect to a reduction of the polaron radius." While Fig. 1 remains illustrative of the crucial qualitative features of the solution, both the (previously imperceptible) negative slope of the bottom of the conduction band and the value of $C_{weak}(0)$ are to be increased, thereby somewhat altering the estimate of the gap between the bands at the transition point.

DYNAMIC PROPERTIES OF A ONE-DIMENSIONAL HEISENBERG MAGNET. S. W. Lovesey and R. A. Meserve [Phys. Rev. Lett. 28, 614 (1972)].

Equation (9) should read

$$\pi F(\vec{k}, \omega) = (1/\chi_{\vec{k}})\, \mathrm{Re} R_{\vec{k}}(i\omega)$$

$$= \frac{\tau\delta_1\delta_2}{[\tau\omega(\omega^2 - \delta_1 - \delta_2)]^2 + (\omega^2 - \delta_1)^2}, \qquad (9)$$

and Eq. (10b) should read

$$\langle\omega^4\rangle = \Omega_0^{\,2}\tfrac{1}{2}\langle\omega^2\rangle\{5 - 3\cos ak + v(1 - 3\cos ak - 3/u) + u[6\cos ak - (2+v)\cos 2ak]\}. \qquad (10b)$$

SOME IMPLICATIONS OF A RECENT TEST OF MUONIC-ATOM THEORY. George A. Rinker, Jr., and Marvin Rich [Phys. Rev. Lett. 28, 640 (1972)].

On page 641, column 2, lines 11 and 12 should read, ••• by the nuclear charge distribution or nuclear polarization uncertainties•••, instead of "••• by the nuclear distribution of nuclear polarization uncertainties."

EVIDENCE FOR LARGE ANTISYMMETRIC SUPEREXCHANGE IN TETRAMERIC COPPER COMPLEXES. M. E. Lines, A. P. Ginsberg, and R. L. Martin [Phys. Rev. Lett. 28, 684 (1972)].

Near the end of the Letter, an omission of a minus sign confuses the final analysis. On page 686, seven lines from the bottom, $J_2 \sim -67$°K should replace the printed $J_2 \sim 67$°K to indicate that it is antiferromagnetic and of opposite sign to J_1.

Reprinted with permission from *Physical Review Letters*, **49**, A. Aspect, J. Dalibard and G. Roger, "Experimental Test of Bell's Inequalities Using Time-Varying Analyzers," pp. 1804-1807, ©1982 American Physical Society.

Experimental Test of Bell's Inequalities Using Time-Varying Analyzers

Alain Aspect, Jean Dalibard,[a] and Gérard Roger
Institut d'Optique Théorique et Appliquée, F-91406 Orsay Cédex, France

(Received 27 September 1982)

Correlations of linear polarizations of pairs of photons have been measured with time-varying analyzers. The analyzer in each leg of the apparatus is an acousto-optical switch followed by two linear polarizers. The switches operate at incommensurate frequencies near 50 MHz. Each analyzer amounts to a polarizer which jumps between two orientations in a time short compared with the photon transit time. The results are in good agreement with quantum mechanical predictions but violate Bell's inequalities by 5 standard deviations.

PACS numbers: 03.65.Bz, 35.80.+s

Bell's inequalities apply to any correlated measurement on two correlated systems. For instance, in the optical version of the Einstein-Podolsky-Rosen-Bohm *Gedankenexperiment*,[1] a source emits pairs of photons (Fig. 1). Measurements of the correlations of linear polarizations are performed on two photons belonging to the same pair. For pairs emitted in suitable states, the correlations are strong. To account for these correlations, Bell[2] considered theories which invoke common properties of both members of the

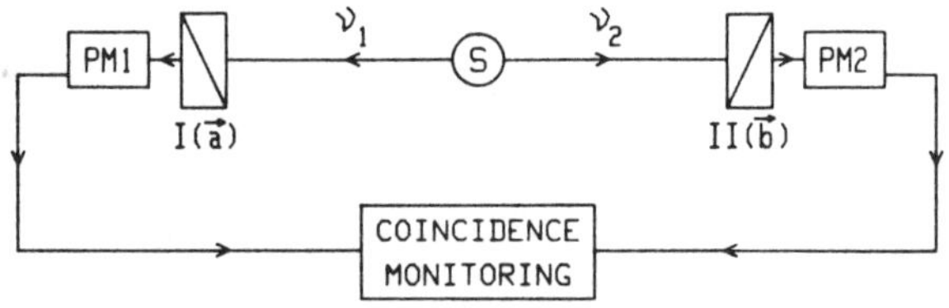

FIG. 1. Optical version of the Einstein-Podolsky-Rosen-Bohm *Gedankenexperiment*. The pair of photons ν_1 and ν_2 is analyzed by linear polarizers I and II (in orientations $\vec{a}$ and $\vec{b}$) and photomultipliers. The coincidence rate is monitored.

VOLUME 49, NUMBER 25 PHYSICAL REVIEW LETTERS 20 DECEMBER 1982

pair. Such properties are referred to as supplementary parameters. This is very different from the quantum mechanical formalism, which does not involve such properties. With the addition of a reasonable locality assumption, Bell showed that such classical-looking theories are constrained by certain inequalities that are not always obeyed by quantum mechanical predictions.

Several experiments of increasing accuracy[3-5] have been performed and clearly favor quantum mechanics. Experiments using pairs of visible photons emitted in atomic radiative cascades seem to achieve a good realization of the ideal *Gedankenexperiment*.[5] However, all these experiments have been performed with static setups, in which polarizers are held fixed for the whole duration of a run. Then, one might question Bell's locality assumption, that states that the results of the measurement by polarizer II does not depend on the orientation $\vec{a}$ of polarizer I (and vice versa), nor does the way in which pairs are emitted depend on $\vec{a}$ or $\vec{b}$. Although highly reasonable, such a locality condition is not prescribed by any fundamental physical law. As pointed out by Bell,[2] it is possible, in such experiments, to reconcile supplementary-parameter theories and the experimentally verified predictions of quantum mechanics: "The settings of the instruments are made sufficiently in advance to allow them to reach some mutual rapport by exchange of signals with velocity less than or equal to that of light." If such interactions existed, Bell's locality condition would no longer hold for static experiments, nor would Bell's inequalities.

Bell thus insisted upon the importance of "experiments of the type proposed by Bohm and Aharonov,[6] in which the settings are changed during the flight of the particles." In such a "timing experiment," the locality condition would then become a consequence of Einstein's causality, preventing any faster-than-light influence.

In this Letter, we report the results of the first experiment using variable polarizers. Following our proposal,[7] we have used a modified scheme (Fig. 2). Each polarizer is replaced by a setup involving a switching device followed by two polarizers in two different orientations: $\vec{a}$ and $\vec{a}'$ on side I, and $\vec{b}$ and $\vec{b}'$ on side II. Such an optical switch is able to rapidly redirect the incident light from one polarizer to the other one. If the two switches work at random and are uncorrelated, it is possible to write generalized Bell's inequalities in a form similar to Clauser-Horne-

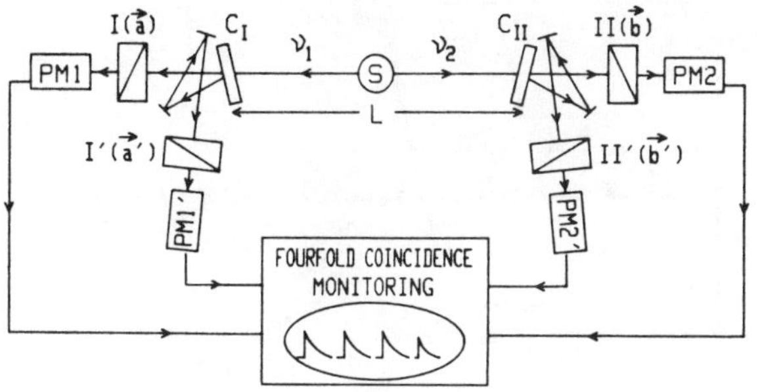

FIG. 2. Timing experiment with optical switches. Each switching device (C_I, C_{II}) is followed by two polarizers in two different orientations. Each combination is equivalent to a polarizer switched fast between two orientations.

Shimony-Holt inequalities[8]:

$$-1 \leqslant S \leqslant 0,$$

with

$$S = \frac{N(\vec{a},\vec{b})}{N(\infty,\infty)} - \frac{N(\vec{a},\vec{b}')}{N(\infty,\infty')} + \frac{N(\vec{a}',\vec{b})}{N(\infty',\infty)} + \frac{N(\vec{a}',\vec{b}')}{N(\infty',\infty')} - \frac{N(\vec{a}',\infty)}{N(\infty',\infty)} - \frac{N(\infty,\vec{b})}{N(\infty,\infty)}.$$

The quantity S involves (i) the four coincidence counting rates [$N(\vec{a},\vec{b})$, $N(\vec{a}',\vec{b})$, etc.] measured in a *single run*; (ii) the four corresponding coincidence rates [$N(\infty,\infty)$, $N(\infty',\infty)$, etc.] with all polarizers removed; and (iii) two coincidence rates [$N(\vec{a}',\infty)$, $N(\infty,\vec{b})$] with a polarizer removed on each side. The measurements (ii) and (iii) are performed in auxiliary runs.

In this experiment, switching between the two channels occurs about each 10 ns. Since this delay, as well as the lifetime of the intermediate level of the cascade (5 ns), is small compared to L/c (40 ns), a detection event on one side and the corresponding change of orientation on the other side are separated by a spacelike interval.

The switching of the light is effected by acousto-optical interaction with an ultrasonic standing wave in water.[9] As sketched in Fig. 3 the incidence angle is equal to the Bragg angle, $\theta_B = 5 \times 10^{-3}$ rad. It follows that light is either transmitted straight ahead or deflected at an angle $2\theta_B$. The light is completely transmitted when the amplitude of the standing wave is null, which occurs twice during an acoustical period. A quarter of a period later, the amplitude of the standing wave is maximum and, for a suitable value of the acoustical power, light is then fully

VOLUME 49, NUMBER 25 PHYSICAL REVIEW LETTERS 20 DECEMBER 1982

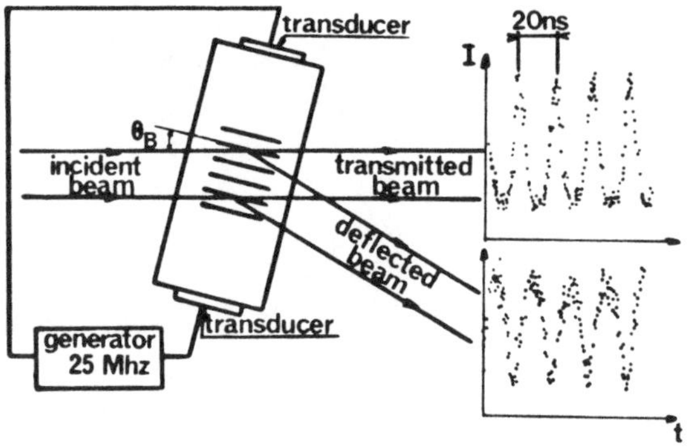

FIG. 3. Optical switch. The incident light is switched at a frequency around 50 MHz by diffraction at the Bragg angle on an ultrasonic standing wave. The intensities of the transmitted and deflected beams as a function of time have been measured with the actual source. The fraction of light directed towards other diffraction orders is negligible.

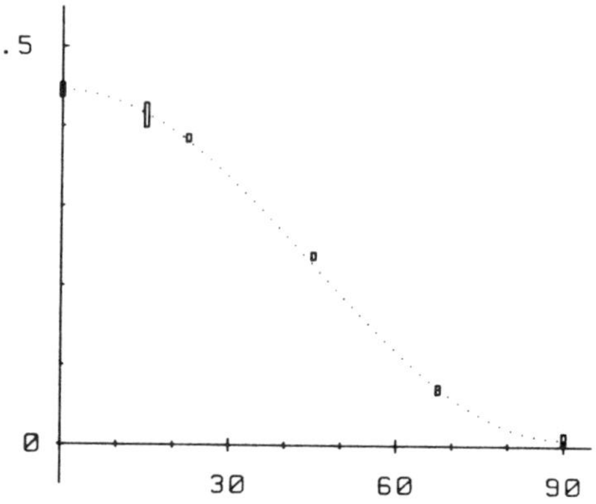

FIG. 4. Average normalized coincidence rate as a function of the relative orientation of the polarizers. Indicated errors are ± 1 standard deviation. The dashed curve is not a fit to the data but the predictions by quantum mechanics for the actual experiment.

deflected. This optical switch thus works at twice the acoustical frequency.

The ultrasonic standing-wave results from interference between counterpropagating acoustic waves produced by two electroacoustical transducers driven in phase at about 25 MHz. In auxiliary tests with a laser beam, the switching has been found complete for an acoustical power about 1 W. In the actual experiment, the light beam has a finite divergence, and the switching is not complete (Fig. 3).

The other parts of the experiment have already been described in previous publications.[4,5] The high-efficiency well-stabilized source of pairs of correlated photons, at wavelengths $\lambda_1 = 422.7$ nm and $\lambda_2 = 551.3$ nm, is obtained by two-photon excitation of a $(J=0)\rightarrow(J=1)\rightarrow(J=0)$ cascade in calcium.

Since each switch is 6 m from the source, rather complicated optics are required to match the beams with the switches and the polarizers. We have carefully checked each channel for no depolarization, by looking for a cosine Malus law when a supplementary polarizer is inserted in front of the source. These auxiliary tests are particularly important for the channels which involve two mirrors inclined at 11°. They also yield the efficiencies of the polarizers, required for the quantum mechanical calculations.

The coincidence counting electronics involve four double-coincidence-counting circuits with coincidence windows of 18 ns. For each relevant pair of photomultipliers, we monitor nondelayed and delayed coincidences. The true coincidence rate (i.e., coincidences due to photons emitted by the same atom) are obtained by subtraction. Simultaneously, a time-to-amplitude converter, followed by a fourfold multichannel analyzer, yields four time-delay spectrums. Here, the true coincidence rate is taken as the signal in the peak of the time-delay spectrum.[4]

We have not been able to achieve collection efficiencies as large as in previous experiments,[4,5] since we had to reduce the divergence of the beams in order to get good switching. Coincidence rates with the polarizers removed were only a few per second, with accidental coincidence rates about one per second.

A typical run lasts 12 000 s, involving totals of 4000 s with polarizers in place at a given set of orientations, 4000 s with all polarizers removed, and 4000 s with one polarizer removed on each side. In order to compensate the effects of systematic drifts, data accumulation was alternated between these three configurations about every 400 s. At the end of each 400-s period, the raw data were stored for subsequent processing with the help of a computer.

At the end of the run, we average the true coincidence rates corresponding to the same configurations for the polarizers. We then compute the relevant ratios for the quantity S. The statistical accuracy is evaluated according to standard statistical methods for photon counting. The processing is performed on both sets of data: that obtained with coincidence circuits, and that obtained with the time-to-amplitude converter. The two methods have always been found to be consistent.

VOLUME 49, NUMBER 25 PHYSICAL REVIEW LETTERS 20 DECEMBER 1982

Two runs have been performed in order to test Bell's inequalities. In each run, we have chosen a set of orientations leading to the greatest predicted conflict between quantum mechanics and Bell's inequalities $[(\vec{a},\vec{b}) = (\vec{b},\vec{a}') = (\vec{a}',\vec{b}') = 22.5°$; $(\vec{a},\vec{b}') = 67.5°]$. The average of the two runs yields

$$S_{\text{expt}} = 0.101 \pm 0.020,$$

violating the inequality $S \leq 0$ by 5 standard deviations. On the other hand, for our solid angles and polarizer efficiencies, quantum mechanics predicts $S_{\text{QM}} = 0.112$.

We have carried out another run with different orientations, for a direct comparison with quantum mechanics. Figure 4 shows that the agreement is excellent.

The new feature of this experiment is that we change the settings of the polarizers, at a rate greater than c/L. The ideal scheme has not been completed since the change is not truly random, but rather quasiperiodic. Nevertheless, the two switches on the two sides are driven by different generators at different frequencies. It is then very natural to assume that they function in an uncorrelated way.

A more ideal experiment with random and complète switching would be necessary for a fully conclusive argument against the whole class of supplementary-parameter theories obeying Einstein's causality. However, our observed violation of Bell's inequalities indicates that the experimental accuracy was good enough for pointing out a hypothetical discrepancy with the predictions of quantum mechanics. No such effect was observed.[10]

This work has been carried out in the group of Professor Imbert, who we thank for his support. We thank all the technical staff of the Institut d'Optique, especially André Villing for the electronics. We are indebted to Jean-Pierre Passérieux, from the Département de Physique Nucléaire et Basse Energie at Centre d'Etudes National de Saclay, for his assistance in fast-coincidence techniques, and to Dr. Torguet and Dr. Gazalé, from Laboratoire d'Acoustooptique de Valenciennes, for help with the optical switches. We also acknowledge the valuable help of Philippe Grangier during the final runs.

[(a)]Permanent address: Laboratoire de Physique de l'Ecole Normale Supérieure, F-75231 Paris Cédex 05, France.

[1]A. Einstein, B. Podolsky, and N. Rosen, Phys. Rev. 47, 777 (1935); D. Bohm, *Quantum Theory* (Prentice-Hall, Englewood Cliffs, N.J., 1951).

[2]J. S. Bell, Physics (N.Y.) 1, 195 (1965).

[3]J. F. Clauser and A. Shimony, Rep. Prog. Phys. 41, 1981 (1978). (This paper is an exhaustive review of the subject); F. M. Pipkin, in *Advances in Atomic and Molecular Physics*, edited by D. R. Bates and B. Bederson (Academic, New York, 1978) (a comprehensive review).

[4]A. Aspect, P. Grangier, and G. Roger, Phys. Rev. Lett. 47, 460 (1981).

[5]A. Aspect, P. Grangier, and G. Roger, Phys. Rev. Lett. 49, 91 (1982).

[6]D. Bohm and Y. Aharonov, Phys. Rev. 108, 1070 (1957). The suggestion of this thought-experiment was already given by D. Bohm, Ref. 1.

[7]A. Aspect, Phys. Lett. 54A, 117 (1975), and Phys. Rev. D 14, 1944 (1976).

[8]J. F. Clauser, M. A. Horne, A. Shimony, and R. A. Holt, Phys. Rev. Lett. 23, 880 (1969).

[9]A. Yariv, *Quantum Electronics* (Wiley, New York, 1975).

[10]Let us emphasize that such results cannot be taken as providing the possiblity of faster-than-light communication. See, for instance, A. Aspect, J. Phys. (Paris), Colloq. 42, C263 (1981).

Cryptographic approach to hidden variables

Reprinted from the *American Journal of Physics,* 43, N. Herbert, "Cryptographic Approach to Hidden Variables," pp. 315-316,

Nick Herbert
Institute for the Study of Consciousness
Berkeley, California 94705
(Received 8 June 1973; revised 18 September 1974)

An elementary proof of Bell's theorem on the incompatibility of local hidden-variable theories with quantum mechanics is presented. The proof is based on considerations of error rates in binary message sequences.

Quantum mechanics does not in general predict the behavior of individual events but only describes the statistical behavior of similarly prepared systems. Whether there is an invisible dynamics beneath quantum mechanics that controls the occurrence of each singular event is the concern of hidden-variable theorists. A successful hidden-variable theory would form a conceptual basis for quantum mechanics in somewhat the same way as kinetic theory undergirds thermodynamics. At first glance it would appear as though any statistical description could easily be given an *ad hoc* hidden-variable underpinning, but quantum mechanics is a peculiar statistical theory and has steadfastly resisted such deeper reformulation.[1]

Interest has recently been revived in these questions by the work of J. S. Bell[2-4] who has shown that for the description of a simple two-particle spin-correlation experiment no *local* hidden-variable theory is consistent with quantum-mechanical predictions. Bell's theorem is in the form of an inequality that must be satisfied if local hidden variables underlie the system. Quantum-mechanical predictions do not satisfy this inequality. It is the purpose of this article to present an elementary derivation of Bell's theorem.

The physical setup is as follows: a spin-zero system (perhaps the $^1S_0$ state of positronium) decays into two photons which travel in opposite directions where they are detected by polarization-sensitive detectors. The two detectors (A and B) are designed to "sort photons" into two bins according to whether they are polarized along the reference axis (A_1,B_1) or perpendicular to it (A_2,B_2). (A Wollaston prism could perform this sorting for optical photons.)

The polarization state of each detected photon is construed to be a "binary message" sent to each detector according to the following scheme: Let the message bit at detector A be a "1" if photon is detected in bin A_1; call it "0" if the photon is detected in bin A_2. The contrary convention is chosen for detector B. When both detector axes are aligned, because of the spin correlation of simultaneously emitted photons, identical (random) binary messages are received at each detector [Fig. 1(a)].

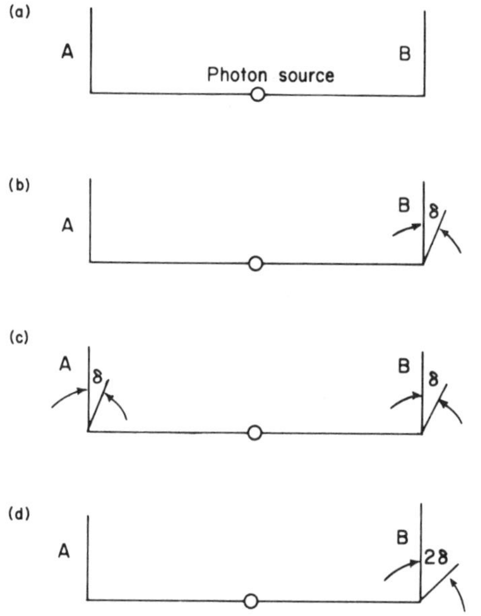

Fig. 1. Special alignments of the "double-message" system: (a) perfect alignment—message at A the same as message at B; (b) B misaligned—$n(\delta)$ error rate introduced in B; (c) A realigned—$n(\delta)$ compensating error rate introduced in A, message at A is now the same as message at B; and (d) B misaligned by 2δ—what is the new error rate?

Suppose the axis of detector B is misaligned by an angle δ. This introduces an error rate $n(\delta)$ into the message stream at B [Fig. 1(b)]. If now detector A is realigned by the same amount δ, the same error rate $n(\delta)$ is introduced into channel A. Moreover the "errors" fall at precisely the correct places in message A to make the two messages again coincide [Fig. 1(c)].

Beginning with the detectors in alignment, change the axis of B by 2δ. What is the new error rate $n(2\delta)$?

Locality assumption: Assuming that changes in the message at B depend only on changes in the setting at B, not on the position of detector A, a change δ in alignment always introduces $n(\delta)$ errors into the B code. But in two successive misalignments, some of these errors may cancel so $n(2\delta)$ can never be greater than $2n(\delta)$:

$$n(2\delta) \leqslant 2n(\delta) \qquad \text{local code production.} \qquad (1)$$

Equation (1) is an inequality that must be satisfied by error rates if the locality assumption is valid. It is a special case of Bell's inequality.

For back-to-back photons, the counting correlation $C(\delta)$ as a function of detector mutual alignment angle δ predicted by quantum mechanics is the same as Malus's $\cos^2\theta$ law for crossed polarizers.

$$C(\delta) = \cos^2\delta. \qquad (2)$$

The error rate is given by

$$n(\delta) = 1 - C(\delta) = 1 - \cos^2\delta$$

$$\text{(quantum mechanics).}^5 \qquad (3)$$

For small δ, this becomes

$$n(\delta) \doteq \delta^2 + O(\delta^4), \quad (4)$$

$$n(2\delta) \doteq 4\delta^2 + O(\delta^4), \quad (5)$$

which does not satisfy the inequality of Eq. (1).

Hence any hidden-variable theory which attempts to explain the occurrence of the binary events at B and the changes in these events as detector B is tilted must take into account the setting of the distant detector A, i.e., the theory must be non-local. This is the striking content of Bell's theorem: No local hidden-variable theory is consistent with quantum mechanics.

It might be supposed, in this cryptographic context, that since the setting at A affects the error rate at B, this effect might be exploited to send faster-than-light messages from A to B. However, if quantum mechanics is really an exhaustive theory, it must be so that the "messages" at B are always perfectly random no matter what the settings of A or B are and no intelligence could ever be decoded from such a featureless sequence.[6]

But if a (necessarily non-local) hidden-variable structure exists beneath quantum phenomena, such super-fast coding is not ruled out. The relativistic impossibility of supraluminal telegraphy may be considered an additional argument against a hidden-variable interpretation of quantum mechanics but the technological advantages of such a rapid communication device seem to make investigations of hidden-variable theories of more than philosophical interest.

[1] A state-of-the-art review of hidden-variable theories is given in F. J. Belinfante, *A Survey of Hidden-Variables Theories* (Pergamon, New York, 1973).

[2] J. S. Bell, Physics **1**, 195 (1965).

[3] E. P. Wigner, Am. J. Phys. **38**, 1005 (1970).

[4] J. F. Clauser, M. A. Horne, A. Shimony, and R. A. Holt, Phys. Rev. Lett. **23**, 880 (1969).

[5] A recent experiment using optical photons gives results consistent with quantum-mechanical predictions. See S. J. Freedman and J. F. Clauser, Phys. Rev. Lett. **28**, 938 (1972).

[6] The random nature of sequences of individual observations cannot be deduced from the customary assumptions of quantum mechanics but must be separately postulated. For details see P. A. Benioff, J. Math. Phys. **11**, 2553 (1970); **12**, 361 (1971).

PHYSICS TERMS MADE EASY

Commutator	A student who drives to school
Corona	An officer who enquires into the manner of violent death
Flux	Past participle of the verb "to flex"
Gram	To review for examinations
Grand canonical ensemble	Ecumenical council
Graph	Principal item of a cow's diet
Ground state	Coffee, before brewing
Harmonic function	Concert
Hermitian operator	Recluse surgeon
Millimeter	A bug like a centimeter, but with more legs
Poynting vector	A redundant term; all vectors point
Spectra	A female ghost
Spin operator	Owner of a Ferris wheel
Ultraviolet catastrophe	Bad sunburn
Watt	Will you please repeat that remark?

—Mostly taken from a "Glossary of Optical Terminology" in a volume presented to Dr. Rudolf Kingslake on his retirement from the Eastman Kodak Company

Reprinted from the *American Journal of Physics,* **46**, A. Peres, "Unperformed Experiments Have No Results," pp. 745-747, ©1978 American Association of Physics Teachers.

Unperformed experiments have no results

Asher Peres

Department of Physics and Atmospheric Science, Drexel University, Philadelphia, Pennsylvania 19104 and Department of Physics, Technion–Israel Institute of Technology, Haifa, Israel[a]

(Received 11 May 1977; accepted 15 January 1978)

This paper discusses correlations between the results of measurements performed on physical systems which are widely separated, but have interacted in the past. It is shown that quantum correlations are *stronger* than classical correlations. This property leads to the following paradox, known as Bell's theorem: Let us assume that the outcome of an experiment performed on one of the systems is independent of the choice of the experiment performed on the other. Now, let us try to imagine the results of alternative measurements, which could have been performed on the same systems *instead* of the actual measurements. Then there is no way of contriving these hypothetical results so that they will satisfy all the quantum correlations with the results of the actual measurements. However, the weaker classical correlations can be satisfied.

I. INTRODUCTION

Many experiments[1] have recently been performed to test predictions of quantum mechanics violating Bell's inequality.[2] The purpose of this paper is twofold. First, we give a simple proof of Bell's inequality, based on purely phenomenological arguments (no mention is made of "hidden variables" or similar superstitions). This inequality can be tested in experiments involving classical correlations or quantum correlations. It is shown that classical correlations respect Bell's inequality, but quantum correlations are *stronger* and lead to its experimental violation.

We then discuss the meaning of this result, again in purely phenomenological terms. The discussion involves a comparison of the results of experiments which were actually performed, with those of hypothetical experiments which could have been performed but were not. It is shown that it is *impossible to imagine* the latter results in a way compatible with (a) the results of the actually performed experiments, (b) long range separability of results of individual measurements, and (c) quantum mechanics.

In the Appendix, we show that a Bell inequality can be constructed (and is violated by quantum mechanics) for any pair of states of any two dynamical variables. However, for the sake of conciseness, the main part of this paper will deal with the simple case of a pair of correlated spins or, classically, of correlated macroscopic angular momenta.

II. CORRELATED DYNAMICAL VARIABLES

Consider a bomb, initially at rest, which explodes into two asymmetric parts carrying angular momenta $\mathbf{J}_1$ and $\mathbf{J}_2 = -\mathbf{J}_1$ (see Fig. 1). An observer detects the first fragment and measures

$$r_\alpha = \mathrm{sign}(\alpha \cdot \mathbf{J}_1),$$

where α is a unit vector in an arbitrary direction, chosen by that observer. Likewise, a second observer detects the other fragment and measures

$$r_\beta = \mathrm{sign}(\beta \cdot \mathbf{J}_2),$$

where β is another unit vector, chosen by the second observer.

Now assume that the directions of $\mathbf{J}_1$ and $\mathbf{J}_2$ are unpredictable and randomly distributed. Then, if the experiment is repeated N times, the averages

$$\langle r_\alpha \rangle = (1/N) \sum_j r_{j\alpha}$$

and

$$\langle r_\beta \rangle = (1/N) \sum_j r_{j\beta}$$

are both close to zero (typically, they are of the order of $1/\sqrt{n}$). Here, $r_{j\alpha}$ and $r_{j\beta}$ ($j = 1, ..., N$) denote r_α and r_β of the jth bomb. We shall henceforth omit the subscript j for brevity.

On the other hand, if the observers compare their results after the latter have been obtained, the average

$$\langle r_\alpha r_\beta \rangle = (1/N) \sum r_\alpha r_\beta$$

may differ from zero. For instance, if $\alpha = \beta$, we shall always have $r_\alpha = -r_\beta$ so that $\langle r_\alpha r_\beta \rangle = -1$. Quite generally, if we consider a unit sphere cut by an equatorial plane perpendicular to α, we shall have $r_\alpha = 1$ if $\mathbf{J}_1$ points through one of the hemispheres and $r_\alpha = -1$ if it points through the other. Likewise a second equatorial plane, perpendicular to β determines the regions with $r_\beta = \pm 1$. The unit sphere is thereby divided by these two equatorial planes into four parts, with $r_\alpha r_\beta = \pm 1$. Their areas are in the ratio $\theta : \pi - \theta$, where θ is the angle between α and β. If $\mathbf{J}_1$ is randomly distributed, we obtain, for large N,

$$\langle r_\alpha r_\beta \rangle = [\theta - (\pi - \theta)]/\pi = -1 + 2\,\theta/\pi.$$

Next, consider two spin-1/2 particles in a singlet state, flying apart from each other in a way that spin is conserved. Our observers now measure $r_\alpha = 2\alpha \cdot \mathbf{s}_1$ and $r_\beta = 2\beta \cdot \mathbf{s}_2$. As in the previous case, r_α and r_β are ± 1, and $\langle r_\alpha \rangle$ and $\langle r_\beta \rangle$ tend to zero for large N, but $\langle r_\alpha r_\beta \rangle$ will not in general be zero: quantum mechanics tells us that, for large N, we should find

$$\langle r_\alpha r_\beta \rangle = \langle \psi | 2\alpha \cdot \mathbf{s}_1 \quad 2\beta \cdot \mathbf{s}_2 | \Psi \rangle.$$

In the singlet state, $\mathbf{s}_2 | \Psi \rangle = -\mathbf{s}_1 | \Psi |$, and as

$$[(-2\alpha \cdot \mathbf{s}_1)(2\beta \cdot \mathbf{s}_1) = (-\alpha \cdot \sigma)(\beta \cdot \sigma) = -\alpha \cdot \beta,$$

we obtain

$$\langle r_\alpha r_\beta \rangle = -\alpha \cdot \beta = -\cos\theta.$$

Figure 2 shows the classical and the quantum correlations. We shall denote both of them as $C(\alpha,\beta)$. Note that

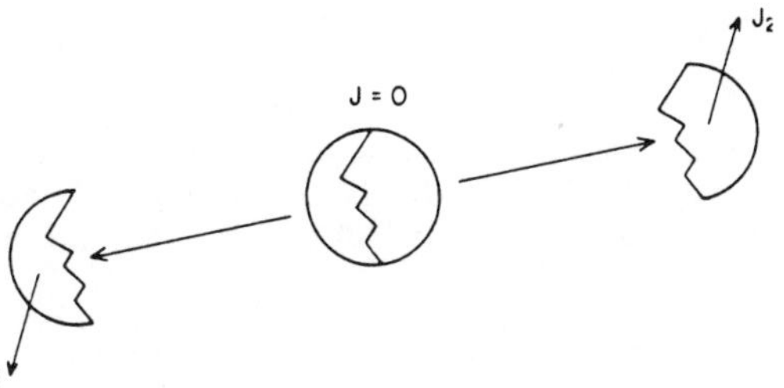

Fig. 1. A bomb, initially at rest, explodes into two fragments carrying opposite angular momenta.

$$|C(\alpha,\beta)_{\text{quantum}}| \geqslant |C(\alpha,\beta)_{\text{classical}}|.$$

This fact has far reaching consequences.

III. BELL'S INEQUALITY

If the observers A and B are far away from each other, we expect intuitively that the results obtained by A are independent of what B is doing and vice versa. For instance, if B had oriented his apparatus along some other direction β', the results obtained by A would have been the same: Not only $\langle r_\alpha \rangle$ would have remained zero, but *each individual* $r_{j\alpha} (j = 1, \cdots, N)$ would have remained unchanged. Indeed, the bomb fragments or the spin-1/2 particles may have started their flight *before* A and B orient their apparatuses along α and β. It thus seems reasonable to assume that B's choice of β cannot affect the result obtained by A, and vice versa.

Let us now try to imagine[3] what could have been the results of measurements by A and B along other directions α' and β'. It is of course impossible to know for sure these results (unless $\alpha' = \pm\alpha$ or $\pm\beta$, etc.) because the experiments were not performed. However, they *could* have been performed (instead of the measurements along α and/or β) and the results could have been only $r_{\alpha'} = \pm 1$ and $r_{\beta'} = \pm 1$.

We can therefore construct a table including both the actual and the hypothetical results (see Table I). We do not know how to fill the last two rows in this table. However, there are only 4^N different ways of doing this, that is, there are only 4^N different tables which can be imagined. We shall soon see that *none* of them may be acceptable.

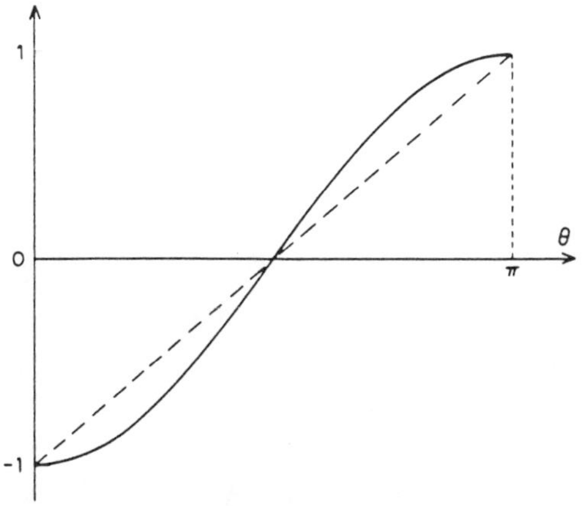

Fig. 2. The quantum correlation (solid line) and the classical one (broken line).

Table I. Comparison of actual and hypothetical measurements.

	Exp. No.	1	2	3	·	·	·	N
Experiments actually performed	$r_\alpha =$	+	+	−	·	·	·	·
	$r_\beta =$	−	+	−	·	·	·	·
Alternative unperformed experiments	$r_{\alpha'} =$	?	?	?	·	·	·	·
	$r_{\beta'} =$	?	?	?	·	·	·	·

To be acceptable, a table must have (for large enough N) $\langle r_{\alpha'} \rangle = \langle r_{\beta'} \rangle = 0$ and moreover

$$\langle r_\alpha r_{\beta'} \rangle = C(\alpha,\beta'),$$
$$\langle r_{\alpha'} r_\beta \rangle = C(\alpha',\beta),$$
$$\langle r_{\alpha'} r_{\beta'} \rangle = C(\alpha',\beta'),$$

because the unperformed experiments *could* have been performed, and then should have given results consistent with the known laws of physics.

Now it is not difficult to see that

$$r_\alpha r_\beta + r_\alpha r_{\beta'} + r_{\alpha'} r_\beta - r_{\alpha'} r_{\beta'} \equiv \pm 2,$$

for any choice of $r_\alpha, \cdots, r_{\beta'}$ (all ± 1). Therefore

$$(1/N) |\textstyle\sum (r_\alpha r_\beta + r_\alpha r_{\beta'} + r_{\alpha'} r_\beta - r_{\alpha'} r_{\beta'})| \leqslant 2.$$

But the left-hand side of this expression is just $|\langle r_\alpha r_\beta \rangle + \langle r_\alpha r_{\beta'} \rangle + \langle r_{\alpha'} r_\beta \rangle - \langle r_{\alpha'} r_{\beta'} \rangle|$. It follows that

$$|C(\alpha,\beta) + C(\alpha,\beta') + C(\alpha',\beta) - C(\alpha',\beta')| \leqslant 2.$$

This is Bell's inequality.

It is easy to show that the latter is violated by the quantum correlation function $C(\alpha,\beta) = -\alpha\cdot\beta$. The maximum violation occurs when all the directions are coplanar, with the (α',β') angle equal to 135° and the three others to 45°. Then

$$\alpha' \cdot \beta = \beta \cdot \alpha = \alpha \cdot \beta' = -\alpha' \cdot \beta' = 1/\sqrt{2},$$

and the left-hand side of Bell's inequality is $2\sqrt{2}$.

Actually, it is more instructive[4] to take $\alpha = \beta$, with α' and β' at an angle θ on each side, so that $\alpha\cdot\beta' = \alpha'\cdot\beta = \cos\theta$ and $\alpha'\cdot\beta' = \cos 2\theta$. The left-hand side of Bell's inequality becomes

$$|-1 - 2\cos\theta + \cos 2\theta| = |-2 - 2\cos\theta(1 - \cos\theta)|$$

which is larger than 2 for any $\theta < 90°$.

On the other hand, if we take the *classical* correlation function $-1 + 2\theta/\pi$, instead of $-\cos\theta$, we get

$$-1 - 2(1 - 2\theta/\pi) + (1 - 4\theta/\pi) = -2$$

exactly. The difference between these results is due to the fact that for small θ, the classical correlation function *falls faster* than the quantum-mechanical one.

IV. DISCUSSION AND OUTLOOK

There are two possible attitudes in the face of these results. One is to say that it is illegitimate to speculate about unperformed experiments. In brief "Thou shalt not think." Physics is then free from many epistemological difficulties. For instance, it is not possible to formulate the EPR paradox.[5]

Alternatively, for those who cannot refrain from thinking, we can abandon the assumption that the results of measurements by A are independent of what is being done by B. In other words, we should not have written $r_{j\alpha}$, but $r_{j\alpha(\beta)}$ or rather $r_{j\alpha(\beta\gamma\delta\cdots)}$ where the additional subscripts $\gamma,\delta,\cdots$ refer to possible experiments on anything else in the world which has been correlated with the history of the jth particle.

This conclusion is surprising. Physicists are used to thinking in terms of isolated systems whose behavior is independent of what happens in the rest of the world (contrary to social scientists, who cannot isolate the subject of their study from its environment). Bell's theorem tells us that such a separation is impossible for individual experiments, although it still holds for averages:

$$(1/N)\sum_j r_{j\alpha(\beta\gamma\delta\cdots)} = 0$$

and

$$(1/N)\sum_j r_{j\alpha(\beta\gamma\delta\cdots)} r_{j\beta(\alpha\gamma\delta\cdots)} = -\alpha\cdot\beta$$

independently of γ,δ, etc.

As it often happens, the subtlety of nature beggars the human imagination.

ACKNOWLEDGMENT

I am grateful to Sophia Wechsler for stimulating discussions.

APPENDIX

The reader may have the impression that Bell's theorem is limited to spin correlations (or photon correlations[6] which have the same algebra). In this Appendix, we show that it can easily be generalized to any pair of states of any two suitably correlated systems.

Let $|m\rangle$ and $|n\rangle$ be two orthogonal states of a physical system. Consider the family of operators

$$F(\alpha) = e^{i\alpha}|m\rangle\langle n| + e^{-i\alpha}|n\rangle\langle m|.$$

These operators have eigenvalues ± 1 corresponding to eigenvectors $(e^{i\alpha/2}|m\rangle \pm e^{-i\alpha/2}|n\rangle)/\sqrt{2}$, and 0 corresponding to all states orthogonal to $|m\rangle$ and $|n\rangle$. Thus any measurement of $F(\alpha)$ shall yield the result $r_\alpha = \pm 1$ or 0. Note that

$$[F(\alpha), F(\alpha')] = 2i\sin(\alpha-\alpha')(|m\rangle\langle m| - |n\rangle\langle n|)$$

does not vanish (unless $\alpha-\alpha'$ is a multiple of π) so that it is impossible to measure both operators simultaneously.

Likewise, let $|\mu\rangle$ and $|\nu\rangle$ be two orthogonal states of another physical system and let us consider the family of operators

$$\Phi(\beta) = e^{i\beta}|\mu\rangle\langle\nu| + e^{-i\beta}|\nu\rangle\langle\mu|.$$

Any measurement of $\Phi(\beta)$ shall yield the result $r_\beta = \pm 1$ or 0. Now, let

$$|\Psi\rangle = (1/\sqrt{2})\,(|m\rangle|\nu\rangle - |n\rangle|\mu\rangle)$$

be the wave function describing an ensemble of pairs of these systems. The elements of each pair can be well separated from each other, yet they are correlated. A straightforward calculation gives

$$\langle\Psi|F(\alpha)\Phi(\beta)|\Psi\rangle = -\cos(\alpha-\beta),$$

and this must be the limiting value of $N^{-1}\sum(r_\alpha r_\beta)$ for large N.

On the other hand, if we consider the possibility of measuring $F(\alpha')$ and/or $\Phi(\beta')$ instead of $F(\alpha)$ and/or $\Phi(\beta)$ we get, as before,

$$|r_\alpha r_\beta + r_\alpha r_{\beta'} + r_{\alpha'} r_\beta - r_{\alpha'} r_{\beta'}| \leqslant 2.$$

Indeed, if none of $r_\alpha,\cdots,r_{\beta'}$ is zero, this is exactly 2. If one or more is zero, the result can also be 1 or 0. So, we again obtain Bell's inequality, which is violated by the above cosine correlation.

[a]Permanent address.

[1]J. F. Clauser, Phys. Rev. Lett. **36**, 1223 (1976); E. S. Fry and R. C. Thompson, Phys. Rev. Lett. **37**, 465 (1976); A. Aspect, Phys. Rev. D **14**, 1944 (1976); M. Lamehi-Rachti and W. Mittig, Phys. Rev. D **14**, 2543 (1976).

[2]J. Bell, Physics (L. I. City, NY) **1**, 195 (1964–65).

[3]H. P. Stapp, Phys. Rev. D **3**, 1303 (1971); P. H. Eberhard, Nuovo Cimento B **38**, 75 (1977).

[4]N. Herbert, Am. J. Phys. **43**, 315 (1975).

[5]A. Einstein, B. Podolsky, and N. Rosen, Phys. Rev. **47**, 777 (1935). The key point in the EPR argument is on p. 779: "If ··· we had chosen another quantity ··· we should have obtained ···."

[6]J. M. Jauch and F. Rohrlich, *The Theory of Photons and Electrons* (Addison-Wesley, Cambridge, MA, 1955), p. 45.

Reprinted from *Physics Review,* **D10**, J.F. Clauser and M.A. Horne, "Experimental Consequences of Objective Local Theories," pp. 526-535, ©1974 American Physical Society.

PHYSICAL REVIEW D VOLUME 10, NUMBER 2 15 JULY 1974

Experimental consequences of objective local theories*

John F. Clauser

Department of Physics and Lawrence Berkeley Laboratory, University of California, Berkeley, California 94720

Michael A. Horne

Department of Physics, Stonehill College, North Easton, Massachusetts 02356

(Received 10 August 1973; revised manuscript received 8 April 1974)

A broad class of theories, called "objective local theories," is defined, motivation for considering these theories is given, and experimental consequences of the class are investigated. An extension of previous analyses by Bell and by Clauser *et al.* shows that predictions of objective local theories and of quantum mechanics differ, and that an experimental test of the entire family of objective local theories can be performed. The experimental requirements are given. Objective local theories satisfying a plausible but experimentally untestable supplementary assumption are shown to be incompatible with existing experimental data.

I. INTRODUCTION

Two papers by Bell have shown that the statistical predictions of quantum mechanics, for certain spatially separated yet correlated two-particle systems, are incompatible with a broad class of local theories. Bell's earlier paper considers the consequences of a physically reasonable locality condition within the domain of an ideal *Gedankenexperiment*.[1] He demonstrates that any theory which satisfies the locality condition must also be deterministic if certain quantum-mechanical predictions are valid for the idealized case. Bell's further analysis shows, however, that any deterministic local theories are necessarily incompatible with some other quantum-mechanical predictions for the *Gedankenexperiment*.

Upon examining the proof in Bell's earlier paper, one might conjecture that it is essentially the deterministic character of the class of theories that is incompatible with quantum mechanics.[2] That is, if the hypotheses assumed for the *Gedankenexperiment* were slightly relaxed so that determinism is no longer derivable,[3] then the incompatibility with the quantum-mechanical predictions will also be removed. This conjecture is incorrect. Bell shows in his second paper[4] that any stochastic theory satisfying the locality condition is also incompatible with quantum mechanics.

Inspired by Bell's first paper, Clauser *et al.*[5] showed that his analysis can be extended to cover realizable systems and that experimental tests of the broad class of local theories covered by these theorems can be performed. Although existing two-particle sources and/or analyzer-detector apparatuses appear insufficiently close to ideal for the desired experiment, Clauser *et al.* showed that, with a plausible but untestable supplementary assumption concerning detector efficiencies, deterministic local theories are incompatible with the quantum-mechanical predictions for a realizable experiment. Experimental results obtained by Freedman and Clauser[6] are in excellent agreement with the relevant quantum-mechanical predictions and thereby indicate that any deterministic local theory is untenable if the supplementary assumption is true. Clauser *et al.* in their original proposal for this experiment restricted their discussion to deterministic local theories. Following Bell's second paper, however, it was noted[7] that the experiment also indicates that any stochastic local theory is untenable if the same supplementary assumption concerning the detectors is made. However, the question of the experimental testability of stochastic local theories with either a weaker auxiliary assumption or with no auxiliary assumption has not previously been discussed.

The present paper extends the previous discussions of deterministic and stochastic local theories in several respects: (a) In preparation for the extension, we characterize explicitly a broad class of theories, which we designate as *objective local theories* (OLT), and discuss the fundamental premises which motivate them. Incompatibility of this class of theories with quantum mechanics has essentially been demonstrated by Bell,[4] but the result is in a form that is not practically experimentally testable. (b) We give a new incompatibility theorem that yields an experimentally testable result. (c) We show that, without an auxiliary assumption, neither the existing results of Freedman and Clauser nor any future refinement of their experiment employing more efficient detectors can provide a test of OLT because the angular correlation of the photon pairs is unsuitable. We note, however, that there do exist two-particle sources which are suitable for a test. (d) We state a supplementary assumption, weaker than that previously employed, and prove that it is sufficient to ensure the incompatibility of OLT and the experimental results of Freedman and Clauser. (e) We construct an explicit OLT model which reproduces the results of that experiment. We thereby prove that the Freedman-Clauser results constitute a refutation of only those OLT which satisfy our (or some similar) supplementary assumption.

II. OBJECTIVE LOCAL THEORIES

We will formulate and motivate objective local theories in the context of the experimental arrangement shown schematically in Fig. 1. A source of coincident two-particle emissions is viewed by two analyzer-detector assemblies 1 and 2. Each apparatus has an adjustable parameter; let a denote the value of the parameter at apparatus 1, and b that at apparatus 2. In Fig. 1, a and b are taken to be angles specifying the orientations of the analyzers, e.g., the axes of linear polarizers for photons, or the directions of the field gradients of Stern-Gerlach magnets for spin-$\frac{1}{2}$ particles. However, neither of these particular interpretations of the parameters a and b is essential for the discussion which follows; a and b may denote the values of any adjustable parameter at apparatus 1 and 2, respectively. Finally, in addition to an adjustable component and a detector, each apparatus may (and in practice does) contain various other components, such as additional filters to shield the detectors from unwanted radiations, etc. Since we require that these additional apparatus components remain in place during the experiment, we will ignore them in the discussion. Similarly, we ignore and assume constant any other macroscopic variables, such as those describing the source-apparatus geometry.

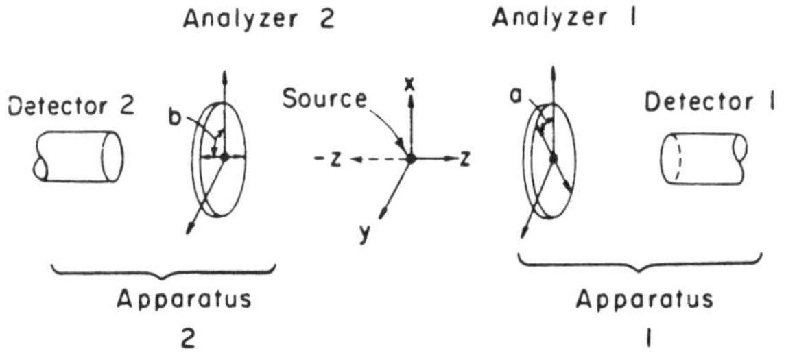

FIG. 1. Scheme considered for a discussion of objective local theories. A source emitting particle pairs is viewed by two apparatuses. Each apparatus consists of an analyzer and an associated detector. The analyzers have parameters, a and b respectively, which are externally adjustable. In the above example, a and b represent the angles between the analyzer axes and a fixed reference axis.

During a period of time, while the adjustable parameters have the values a and b, the source emits, say, N of the two-particle systems of interest.[8] For this period, denote by $N_1(a)$ and $N_2(b)$ the number of counts at detectors 1 and 2, respectively, and by $N_{12}(a, b)$ the number of simultaneous counts from the two detectors (coincident counts).[9] If N is sufficiently large, then the ensemble probabilities of these results are

$$p_1(a) = N_1(a)/N\,,\quad p_2(b) = N_2(b)/N\,,\quad p_{12}(a, b) = N_{12}(a, b)/N\,. \tag{1}$$

Consider one of the two-component emissions from the source. Physical theories, classical, quantum-mechanical, and presumably more general ones as well, characterize a physical system with a *state*. Moreover, during the system's existence, its state in general evolves. Consider the state specification of the above system at a time intermediate between its emission and its impingement on either apparatus.[10] Denote this state by λ. Note that we do not necessarily make a commitment to the completeness of this state specification, i.e., it may or may not describe the ultimate essence of the system at the chosen time. But neither do we make any restriction on the possible complexity of λ, nor do we assume it has any special characteristics; in short, we assume no model. As the state described initially by λ subsequently evolves, it may or may not trigger a count at apparatus 1, and similarly it may or may not do so at apparatus 2. The initial state λ, if it serves the same role as in existing theories, will suffice to determine *at least* the probabilities of these events.[11] Let the probabilities of a count being triggered at apparatus 1 and 2 be $p_1(\lambda, a)$ and $p_2(\lambda, b)$, respectively, and let $p_{12}(\lambda, a, b)$ be the probability that both counts are triggered.[12]

Since, in general, every system in the ensemble emitted by the source may not have the same initial state, we allow a mixture of states. Let $\rho(\lambda)$ be the normalized probability density characterizing the ensemble of emissions.[13] In terms of the quantities just defined, the ensemble probabilities given in Eqs. (1) are

$$p_1(a) = \int_\Gamma d\lambda\, \rho(\lambda) p_1(\lambda, a)\,,\quad p_2(b) = \int_\Gamma d\lambda\, \rho(\lambda) p_2(\lambda, b)\,,\quad p_{12}(a, b) = \int_\Gamma d\lambda\, \rho(\lambda) p_{12}(\lambda, a, b)\,, \tag{2}$$

where Γ is the space of the states λ. The formulation (2) is quite general. Nothing so far has been assumed that is not satisfied by quantum mechanics It suffices in Eqs. (2) to let λ define a quantum state and let $\rho(\lambda)$ define a distribution over quantum states implicit in a mixture.

Hereafter, we focus our attention on a special case of formulation (2) in which

$$p_{12}(\lambda, a, b) = p_1(\lambda, a) p_2(\lambda, b) . \quad (2')$$

What considerations motivate this factored form? Clearly, if each source emission consists of two well-localized subsystems, e.g., a pair of objective particles, and there is no action at a distance, then the factored form is a reasonable locality condition. More generally, the factored form is a natural expression of a field-theoretical point of view, which in turn is an extrapolation from the common-sense view that there is no action at distance.

Fields propagate and can impinge upon different localized objects, among them pieces of apparatus. Since λ describes the field initially,[14] and the parameter a is associated with only one apparatus, it is reasonable that there is a well-defined probability $p_1(\lambda, a)$ that that apparatus will be triggered. The impingement of the field on this apparatus will naturally modify the field. However, if the events marking the action of the field on two pieces of apparatus (triggering or not in each case) have spacelike separation, then the reaction upon the field due to impingement on the first apparatus will not have time to affect the part of the field impinging the second apparatus, and conversely. Hence, the probability p_2 that the second apparatus will be triggered will not depend on whether or not the first has been triggered, or upon the choice of the parameter a at the first apparatus, or even upon the presence of the first apparatus. A similar assertion of independence holds for p_1, whence (2′) follows. In view of the foregoing motivation of Eq. (2′), we call any theory in which it holds an objective local theory.[15]

It should be recalled that the wave function was introduced by Schrödinger from a field-theoretical point of view, and most physicists have continued to think of it field-theoretically, whatever their precautions when they are on guard about the character of quantum mechanics. Hence, we conjecture that Eq. (2′) is implicit in the thinking of many physicists.[16] Whether or not this is correct, it is apparent that quantum mechanics is not of the form (2′). For many two-quanta sources, every emission is described by the same pure state, suggesting that this quantum state be identified with one value of λ, say λ', and that $\rho(\lambda) = \delta(\lambda - \lambda')$ (no mixture). However, when the two-particle quantum state is not a simple product of single-particle states but is instead a superposition of such products, the quantum-mechanical probability p_{12} does not in general admit the factorization (2′). The only alternative for saving Eq. (2′) is to identify the quantum-mechanical pure state with $\rho(\lambda)$, i.e., with a mixture of other states λ.[17] But, as we shall see, this identification is impossible for some quantum-mechanical pure states.

III. EXPERIMENTAL CONSEQUENCES

Measurement of the probabilities (1) requires not only the numbers $N_1(a)$, $N_2(b)$, and $N_{12}(a, b)$, which are directly observed quantities in a counting experiment, but also the number N, which is generally unobservable in an experiment counting microsystems. In practice, the number of emissions N occurring during any time period is usually deduced from the counting data for that period, since intervening counters will in practice depolarize (if not destroy) the systems. But this deduction always depends upon the currently accepted theoretical description of the whole phenomenon—the source, the apparatus, and their interactions. That is, N is actually deduced from Eqs. (1) *themselves* with the p's supplied by the theory at hand. Clearly, any such method of determining N must not be employed in an experimental test of competing theories. Therefore, in this section, we derive a consequence of Eq. (2′) which is experimentally testable without N being known, and which contradicts the quantum-mechanical predictions.

Let a and a' be two orientations of analyzer 1, and let b and b' be two orientations of analyzer 2. The inequalities

$$\begin{aligned} &0 \le p_1(\lambda, a) \le 1 , \\ &0 \le p_1(\lambda, a') \le 1 , \\ &0 \le p_2(\lambda, b) \le 1 , \\ &0 \le p_2(\lambda, b') \le 1 \end{aligned} \quad (3)$$

hold if the probabilities are sensible. These inequalities and the theorem in Appendix A give

$$-1 \le p_1(\lambda, a) p_2(\lambda, b) - p_1(\lambda, a) p_2(\lambda, b') + p_1(\lambda, a') p_2(\lambda, b) + p_1(\lambda, a') p_2(\lambda, b') - p_1(\lambda, a') - p_2(\lambda, b) \le 0$$

for each λ. Multiplication by $\rho(\lambda)$ and integration over λ gives

$$-1 \le p_{12}(a, b) - p_{12}(a, b') + p_{12}(a', b) + p_{12}(a', b') - p_1(a') - p_2(b) \le 0 \quad (4)$$

as a necessary constraint on the statistical predictions of any OLT. If, for some reason such as rotational invariance, it is found experimentally that $p_1(a)$ and $p_2(b)$ are constant, and that $p_{12}(a, b) = p_{12}(\phi)$ holds, where $\phi = |b - a|$ is the angle between the analyzer axes, then (4) becomes

$$-1 \leq 3p_{12}(\phi) - p_{12}(3\phi) - p_1 - p_2 \leq 0 \; . \qquad (4')$$

Here, a, a', b, and b' have been chosen so that

$$|a - b| = |a' - b| = |a' - b| = \tfrac{1}{3}|a - b'| = \phi \; .$$

The upper limits in (4) and in (4′) are experimentally testable without N being known. Inequalities (4) and (4′) hold perfectly generally for any systems described by OLT. These are new results not previously presented elsewhere. The relationship between them and the previous inequalities of Bell is discussed in Appendix B.

IV. INCOMPATIBILITY WITH QUANTUM MECHANICS

We now present the conditions under which the predictions (4) and (4′) are incompatible with those of quantum mechanics. Consider an experiment with a configuration as described above whose quantum-mechanical predictions take the following form:

$$\begin{aligned} p_{12}(\phi) &= \tfrac{1}{4}\eta_1\eta_2 f_1 g[\epsilon_+^1\epsilon_+^2 + \epsilon_-^1\epsilon_-^2 F\cos(n\phi)] \; , \\ p_1 &= \tfrac{1}{2}\eta_1 f_1 \epsilon_+^1 \; , \\ p_2 &= \tfrac{1}{2}\eta_2 f_2 \epsilon_+^2 \; . \end{aligned} \qquad (5)$$

This general form is characteristic of many correlation phenomena, e.g., the spin-$\frac12$–spin-$\frac12$ correlation *Gedankenexperiment* introduced by Bohm[18] and used by Bell,[1,4] and the actual experiments of Freedman and Clauser,[6] Wu and Shaknov,[19] and Kasday *et al.*[20] Inserting the predictions (5) into (4′) and selecting the optimum value $\phi = 45°/n$, one finds that the condition for a violation of the upper bound is

$$\eta g \epsilon_+[\sqrt{2}\,(\epsilon_-/\epsilon_+)^2 F + 1] \geq 2 \; . \qquad (6)$$

Here for simplicity we have taken $\eta_1 = \eta_2 \equiv \eta$, $f_1 = f_2$, $\epsilon_+^1 = \epsilon_+^2 \equiv \epsilon_+$, and $\epsilon_-^1 = \epsilon_-^2 \equiv \epsilon_-$. Thus, a correlation experiment with values in the domain specified by (6) is capable of distinguishing between OLT and quantum theory. Although experiments are possible for which this is the case, there are at present no existing experimental results satisfying (6), and thus none which are in violation of (4′) and/or (4).

Consider next the specific example of the photon pairs emitted by a $J=0 \to J=1 \to J=0$ atomic cascade. A source of such pairs was used in the experiment of Freedman and Clauser. Figure 2 is a diagram of their experiment; it is clearly of the configuration discussed in the previous section. For this arrangement, the predictions are given by Eqs. (5) with $n=2$ and the following identifications[5]: η_i is the quantum efficiency of detector i ($i = 1, 2$), and

$$\epsilon_+^i \equiv \epsilon_M^i + \epsilon_m^i \text{ and } \epsilon_-^i \equiv \epsilon_M^i - \epsilon_m^i \; .$$

Here ϵ_M^i is the efficiency of polarizer i for light polarized parallel to the polarizer axis, and ϵ_m^i is the efficiency for light polarized perpendicular to it. The function $f_1 = f_2 = f(\theta)$ is the probability that the $J=0 \to J=1$ ($J=1 \to J=0$) emission enters apparatus 1 (apparatus 2),

$$f(\theta) = \tfrac{1}{2}(1 - \cos\theta) \; , \qquad (7)$$

with θ being the half-angle of the cones subtended by each detector aperture. The function $g = g(\theta)$ is the conditional probability, or angular correlation factor, that if the $J=0 \to J=1$ emission enters apparatus 1 then the $J=1 \to J=0$ emission will enter apparatus 2,

$$g(\theta) = \frac{3}{8}\,\frac{[G_2(\theta)]^2 + \frac12[G_3(\theta)]^2}{1 - \cos\theta} \; . \qquad (8)$$

Finally,

$$F = F(\theta) = \frac{2[G_1(\theta)]^2}{[G_2(\theta)]^2 + \frac12[G_3(\theta)]^2} \qquad (9)$$

reflects a depolarization effect due to noncollinearity of the two photons and approaches unity for infinitesimal detector apertures ($\theta \to 0$). The functions G_1, G_2, and G_3 are given in Ref. 5.

Inserting Eqs. (8) and (9) into (6) one finds that because of the relatively small value of $g(\theta)$ condition (6) is *not* satisfied for any value of the detector half-angle θ, even if the analyzer and detector efficiencies are ideal ($\eta = \epsilon_+ = \epsilon_- = 1$).[21] Therefore, for cascade-photon experiments, the quantum-mechanical predictions are compatible with (4) even in the domain of ideal apparatus. The

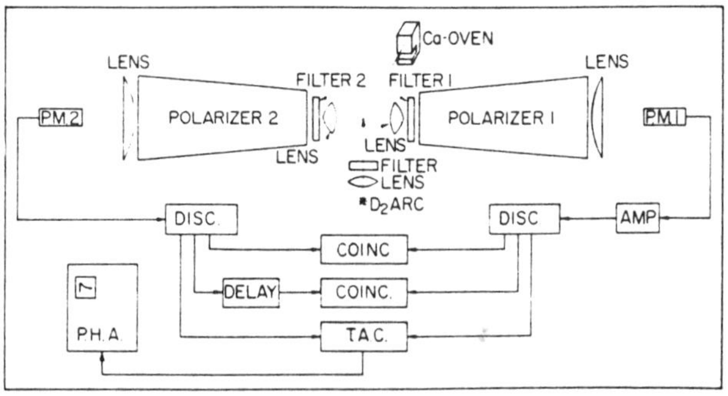

FIG. 2. Schematic diagram of apparatus and associated electronics of the experiment by Freedman and Clauser. Scalers (not shown) monitored the outputs of the discriminators and coincidence circuits. (Figure after Freedman and Clauser.)

insufficient magnitude of the angular correlation factor $g(\theta)$ is a consequence of the fact that an atomic cascade is a three-body decay, the atom being the third body.

However, for correlated particles produced in certain two-body decays, the quantum-mechanical predictions violate (4). The annihilation of ground-state positronium into two γ rays or the dissociation of a spin-0 or a spin-1 molecule into two spin-$\frac{1}{2}$ particles produce correlations of the form (5). Since these are two-body decays, $g(\theta)=1$ holds even for small θ (provided the center-of-mass velocity of the decaying object is sufficiently small). Even with $g=1$, inequality (6) imposes rather stringent conditions on the efficiencies of the analyzers and detectors. But there appears to be no *a priori* reason why such conditions cannot be achieved in practice. This question will be the subject of future work.

V. CONSEQUENCES WITH A SUPPLEMENTARY ASSUMPTION

Until a correlation experiment employing highly efficient analyzers and detectors is performed on two-body decays,[22] it is desirable to exhibit a physically plausible supplementary assumption which makes the existing cascade-photon experiment applicable as a test of OLT. In this section we state an assumption, weaker than the one previously presented by Clauser *et al.*, and prove that it is sufficient to make OLT incompatible with existing experimental results. In the next section we prove, with an explicit OLT model, that this or some other supplementary assumption is necessary for such an application.

The assumption is that, for every emission λ, the probability of a count with a polarizer in place is less than or equal to the probability with the polarizer removed. Let ∞ denote the absence of the polarizer, and let $p_1(\lambda,\infty)$ denote the probability of a count from detector 1 when the polarizer is absent and the emission is λ. A similar probability $p_2(\lambda,\infty)$ may be defined for apparatus 2. Thus, our assumption is

$$0 \leqslant p_1(\lambda, a) \leqslant p_1(\lambda,\infty) \leqslant 1\,, \qquad 0 \leqslant p_2(\lambda, b) \leqslant p_2(\lambda,\infty) \leqslant 1\,, \tag{10}$$

for every λ, and for all values of a and b. We call this the *no-enhancement assumption*.[23] We now make an argument analogous to that which led from (3) to (4). Inequalities (10) and the theorem of Appendix A yield immediately the result

$$-p_{12}(\infty,\infty) \leqslant p_{12}(a,b) - p_{12}(a,b') + p_{12}(a',b) + p_{12}(a',b') - p_{12}(a',\infty) - p_{12}(\infty,b) \leqslant 0\,, \tag{11}$$

where

$$p_{12}(x,y) \equiv \int_\Gamma d\lambda\, \rho(\lambda) p_1(\lambda,x) p_2(\lambda,y)$$

for all x and y. With the same conditions used in writing (4′), (11) becomes

$$p_{12}(\infty,\infty) \leqslant 3p_{12}(\phi) - p_{12}(3\phi) - p_{12}(a',\infty) - p_{12}(\infty,b) \leqslant 0 \tag{11′}$$

Note that all terms in (11) are joint probabilities for coincident counts at the two detectors. Inequality (4), in contrast, contains the two terms p_1 and p_2 which are probabilities of a count at a single detector. The upper limit of (11), or (11′), is identical to the previous result of Clauser *et al.*, but their derivation was restricted to deterministic local theories and employed an auxiliary assumption stronger than the no-enhancement assumption.

The quantum-mechanical prediction for $p_{12}(\phi)$ in the cascade-photon experiment was given in Eqs. (5) and (7)–(9). The predictions for the other joint probabilities occurring in (11) are[5]

$$p_{12}(a',\infty) = \tfrac{1}{2}\eta_1\eta_2 f(\theta) g(\theta)\epsilon_+^1\,, \qquad p_{12}(\infty,b) = \tfrac{1}{2}\eta_1\eta_2 f(\theta) g(\theta)\epsilon_+^2\,, \qquad p_{12}(\infty,\infty) = \eta_1\eta_2 f(\theta) g(\theta)\,. \tag{12}$$

These predictions violate the upper bound in (11′) provided

$$\epsilon_+[\sqrt{2}(\epsilon_-/\epsilon_+)^2 F(\theta) + 1] \geqslant 2 \tag{13}$$

holds. As before, we have used the optimum value $\phi = 22\frac{1}{2}^\circ$ and set $\epsilon_+^1 = \epsilon_+^2 \equiv \epsilon_+$ and $\epsilon_-^1 = \epsilon_-^2 \equiv \epsilon_-$. Note that neither the angular correlation factor nor the detector efficiencies appear in (13). The apparatus of Freedman and Clauser satisfied (13), and the experimental results, which confirmed the quantum-mechanical predictions, substantially violated the upper bound of (11′). Consequently, in view of the theorem just given, any OLT that do not admit enhancement are untenable.[24]

VI. NECESSITY OF A SUPPLEMENTARY ASSUMPTION

To prove that an auxiliary assumption is necessary if the cascade-photon experiment is to refute OLT, we exhibit an explicit OLT model which reproduces the results of that experiment. Since the experimental results are in agreement with the quantum-mechanical predictions, it suffices to construct a model which reproduces these predictions. For simplicity, however, we exhibit the model only for the ideal case in which the detectors subtend infinitesimal solid angles $[\theta \to 0$ and $F(\theta) \to 1]$. The extension of the model to the finite solid angles of the actual experiment $[\theta = 30°$ and $F(\theta) = 0.99]$ introduces nothing new, and the additional complexity obscures the point. With these simplifications, the predictions to reproduce are

$$p_{12}(\phi)/p_{12}(\infty,\infty) = \tfrac{1}{4}(\epsilon_+^1\epsilon_+^2 + \epsilon_-^1\epsilon_-^2 \cos 2\phi)\,,$$
$$p_{12}(a,\infty)/p_{12}(\infty,\infty) = \tfrac{1}{2}\epsilon_+^1\,, \tag{14}$$
$$p_{12}(\infty,b)/p_{12}(\infty,\infty) = \tfrac{1}{2}\epsilon_+^2\,.$$

Only ratios could be measured, since, with N unknown, the actual values of the p's were experimentally inaccessible.

The model is as follows: Each emission pair consists of two particles, such that particle 1 travels along the $+z$ axis to apparatus 1, and particle 2 travels along the $-z$ axis to apparatus 2. Both members of the pair possess a common state variable λ which is simply an azimuthal angle; that is, it specifies a direction perpendicular to the flight axis from some reference axis (see Fig. 3). The ensemble of emitted pairs is characterized by a normalized isotropic density

$$\rho(\lambda)d\lambda = \frac{d\lambda}{2\pi}\,, \quad 0 \le \lambda \le 2\pi\,. \tag{15}$$

From the same reference axis, we specify the orientations of polarizers 1 and 2 by the angles a and b. With the polarizer removed, the probability of a count at each detector is a constant, independent of λ:

$$p_1(\lambda,\infty) = c_1\,, \tag{16}$$
$$p_2(\lambda,\infty) = c_2\,. \tag{17}$$

The probability of a count at detector 1, given an emission λ and the setting a of an inserted polarizer, is

$$p_1(\lambda,a) = \tfrac{1}{2}c_1[\epsilon_+^1 + \epsilon_-^1 \cos 2(\lambda - a)]\,. \tag{18}$$

At detector 2, the probability is

$$p_2(\lambda,b) = \tfrac{1}{2}c_2\pi\epsilon_+^2/\delta \tag{19}$$

for $b - \frac{1}{2}\delta \le \lambda \le b + \frac{1}{2}\delta$ and $\pi + b - \frac{1}{2}\delta \le \lambda \le \pi + b + \frac{1}{2}\delta$; $p_2(\lambda,b)$ is zero otherwise. Here δ is a function of the ratio $\epsilon_-^2/\epsilon_+^2$, and is defined by the relation

$$(\sin\delta)/\delta = \epsilon_-^2/\epsilon_+^2\,. \tag{20}$$

Clearly, this model is an OLT provided the probabilities (16) through (19) are less than unity. Evaluation of the ensemble probabilities yields

$$p_{12}(a,b) = \tfrac{1}{4}c_1c_2[\epsilon_+^1\epsilon_+^2 + \epsilon_-^1\epsilon_-^2 \cos 2(a-b)]\,,$$
$$p_{12}(a,\infty) = \tfrac{1}{2}c_1c_2\epsilon_+^1\,,$$
$$p_{12}(\infty,b) = \tfrac{1}{2}c_1c_2\epsilon_+^2\,,$$
$$p_{12}(\infty,\infty) = c_1c_2\,,$$

which agree with the quantum-mechanical ratios (14).

Note that for some values of the polarizer parameters ϵ_+^i and $\epsilon_-^i/\epsilon_+^i$ the model is enhancement-free. Since the function $p_1(\lambda,a)$ is less than $p_1(\lambda,\infty)$ for all physically sensible values of ϵ_+ and ϵ_-, it is enhancement-free in general. To see this note that, since $0 \le \epsilon_m \le \epsilon_M \le 1$ holds for any polarizer, it follows that $0 \le \epsilon_-^i/\epsilon_+^i \le 1$ and $\epsilon_+^i \le 2[(\epsilon_-^i/\epsilon_+^i)+1]^{-1}$ also hold. However, $p_2(\lambda,b)$ is enhancement-free if and only if

$$\epsilon_+^2 \le 2\delta/\pi\,. \tag{21}$$

The values $\epsilon_+^2 \approx 1.00$ and $\epsilon_-^2/\epsilon_+^2 \approx 0.94$ of the Freedman-Clauser apparatus do not satisfy (21), and therefore, as expected from the theorem of the previous section, the model requires enhancement to reproduce the experimental results. Note finally that, even with enhancement, $p_2(\lambda,b)$ remains sensible provided the constant c_2 of the model is sufficiently small. The value of $p_2(\lambda,b)$ is sensible for all λ and b provided $c_2 \le 2\delta(\pi\epsilon_+^2)^{-1}$. For the actual values of the experiment this condition is $c_2 \le 0.38$.

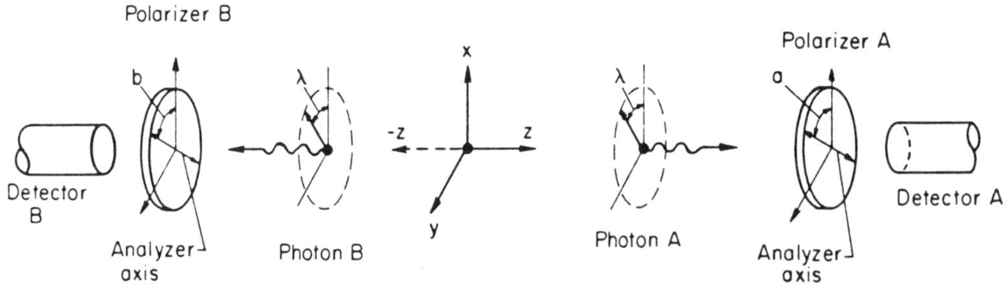

FIG. 3. Coordinate system for OLT model. Photon particles A and B carry the same azimuthal direction λ, which, along with the analyzer orientation a or b, determines the probability of a count at the associated detector.

VII. CONCLUSIONS

Physicists have consistently attempted to model microscopic and macroscopic phenomena in terms of objective entities, preferably with some definable structure. The present paper has addressed the question of whether or not the existing formalism of quantum mechanics can be recast or perhaps reinterpreted in a manner which restores the objectivity of nature, and thus allows such models (deterministic or not) to be made. We have found that it is not possible to do so in a natural way, consistent with locality, without an observable change of the experimental predictions.[26]

ACKNOWLEDGMENT

The authors gratefully acknowledge many valuable discussions with A. Shimony.

APPENDIX A: TWO INEQUALITIES

We prove the following theorem: Given six numbers x_1, x_2, y_1, y_2, X, and Y such that

$$\begin{aligned} &0 \leq x_1 \leq X, \\ &0 \leq x_2 \leq X, \\ &0 \leq y_1 \leq Y, \\ &0 \leq y_2 \leq Y, \end{aligned} \tag{A1}$$

then the function $U = x_1y_1 - x_1y_2 + x_2y_1 + x_2y_2 - Yx_2 - Xy_1$ is constrained by the inequalities

$$-XY \leq U \leq 0 . \tag{A2}$$

To establish the upper bound, consider two cases. First assume that $x_1 \geq x_2$ and rewrite (A2),

$$U = (x_1 - X)y_1 + (y_1 - Y)x_2 + (x_2 - x_1)y_2 .$$

We have thus assumed the last term to be nonpositive. Inequalities (A1) require the first two terms to be nonpositive, and the validity of the upper bound is demonstrated for this case. Next assume the other alternative, i.e., that $x_1 < x_2$, and use this assumption to bound U, thus:

$$\begin{aligned} U &= x_1(y_1 - y_2) + (x_2 - X)y_1 + x_2(y_2 - Y) \\ &\leq x_1(y_1 - y_2) + (x_2 - X)y_1 + x_1(y_2 - Y) \\ &= (x_2 - X)y_1 - x_1(y_1 - Y) \leq 0 . \end{aligned}$$

Thus, the upper bound is established in general.

The proof of the lower bound follows from a consideration of three cases. First, assume $x_2 \geq x_1$. The validity of the lower bound is apparent by inspection when written in the form

$$U + XY = (X - x_2)(Y - y_1) + x_1y_1 + (x_2 - x_1)y_2 ,$$

since (A1) requires all three terms to be nonnegative. Similarly for the case $y_1 \geq y_2$, inspection reveals the correctness of the lower bound when written

$$U + XY = (X - x_2)(Y - y_1) + x_2y_2 + x_1(y_1 - y_2) .$$

Finally, suppose neither of the two previous cases holds; that is, $x_2 < x_1$ *and* $y_1 < y_2$. Then write

$$\begin{aligned} U + XY &= (X - x_2)(Y - y_1) \\ &\quad - (x_1 - x_2)(y_2 - y_1) + x_2 y_1 . \end{aligned}$$

The sum of the first two terms in non-negative since now $(X - x_2) \geq (x_1 - x_2) > 0$ and $(Y - y_1) \geq (y_2 - y_1) > 0$. By (A1) the final term is also nonnegative. Q.E.D.

APPENDIX B: INEQUALITIES (4) AND BELL'S THEOREM

Inequalities (4) assumed the arrangement of Fig. 1, where each apparatus consists of an analyzer with a single-channel output followed by a photomultiplier. Bell[4] considers a two-channel apparatus, e.g., a birefringent polarizing crystal followed by two photomultipliers, one photomuliplier monitoring the ordinary ray emerging from the crystal and the other monitoring the extraordinary ray.[27] Let $p_1^+(\lambda, a)$ denote the probability of a count in the ordinary channel of apparatus 1, and $p_1^-(\lambda, a)$ denote the probability of a count in the extraordinary channel, for the orientation a of the analyzer and initial state λ of the emission. Let $p_2^+(\lambda, b)$ and $p_2^-(\lambda, b)$ be similarly defined for apparatus 2. Bell considers the correlation function defined by[28]

$$\begin{aligned} P(a, b) &\equiv \int_\Gamma d\lambda\, \rho(\lambda)[p_1^+(\lambda, a) - p_1^-(\lambda, a)] \\ &\qquad \times [p_2^+(\lambda, b) - p_2^-(\lambda, b)] \\ &= p_{12}^{++}(a, b) - p_{12}^{+-}(a, b) \\ &\qquad - p_{12}^{-+}(a, b) + p_{12}^{--}(a, b) , \\ p_{12}^{jk}(a, b) &\equiv \int_\Gamma d\lambda\, \rho(\lambda) p_1^j(\lambda, a) p_2^k(\lambda, b) , \\ &\qquad\qquad \text{for } j, k = \pm 1 . \end{aligned} \tag{B1}$$

Using the fact that each square bracket in (B1) is bounded by ± 1, he proves that P is constrained by the inequalities

$$-2 \leq P(a, b) - P(a, b') + P(a', b) + P(a', b') \leq 2 . \tag{B2}$$

Note that for a direct experimental test of (B2) the number of emissions N must be known, and that in actual practice this probably cannot be found without either destroying or at least depolarizing the particles. [If N can be found, (B2) suffices for an experimental test.] Here we wish to show the relations between (4) and (B2).

First, (B2) is a corollary of (4). In an experiment employing two detectors (+ and −) behind each double-channel analyzer, inequalities (4) are still applicable and provide four sets of inequalities,

$$-1 \le p_{12}^{jk}(a,b) - p_{12}^{jk}(a,b') + p_{12}^{jk}(a',b) + p_{12}^{jk}(a',b') - p_1^j(a') - p_2^k(b) \le 0\,, \tag{B3}$$

where $j=\pm 1$ and $k=\pm 1$ indicate which detectors are considered. Multiplying the inequalities for which $j\ne k$ by −1 and adding these to the inequalities for which $j=k$, we obtain (B2).

Second, with Bell's formulation, (4) is not an immediate corollary of (B2). For a given analyzer setting a and emission λ, there are three possible results at apparatus 1: a count in the + detector, a count in the − detector, or no count in either detector. Let $p_1^0(\lambda,a)$ denote the probability of no count. Clearly

$$p_1^+(\lambda,a) + p_1^-(\lambda,a) + p_1^0(\lambda,a) = 1$$

holds, which implies

$$p_1^+(\lambda,a) - p_1^-(\lambda,a) = 2p_1^+(\lambda,a) + p_1^0(\lambda,a) - 1\,. \tag{B4}$$

With (B4), and a similar expression for apparatus 2, the expression (B1) defining the correlation function becomes

$$\begin{aligned} P(a,b) = {} & 4p_{12}^{++}(a,b) - 2p_1^+(a) - 2 - p_2^+(b) + 1 \\ & + 2p_{12}^{+0}(a,b) + 2p_{12}^{0+}(a,b) + p_{12}^{00}(a,b) \\ & - p_1^0(a) - p_2^0(b)\,. \end{aligned} \tag{B5}$$

Insertion of (B5), and similar expressions for $P(a',b)$, $P(a,b')$ and $P(a',b')$, into (B2) yields

$$-1 - Q \le p_{12}^{++}(a,b) - p_{12}^{++}(a,b') + p_{12}^{++}(a',b) + p_{12}^{++}(a',b') - p_1^+(a') - p_2^+(b) \le -Q\,, \tag{B6}$$

where

$$\begin{aligned} Q \equiv \tfrac{1}{2}[& p_{12}^{+0}(a,b) + p_{12}^{0+}(a,b) + \tfrac{1}{2}p_{12}^{00}(a,b) - p_{12}^{+0}(a,b') - p_{12}^{0+}(a,b') - \tfrac{1}{2}p_{12}^{00}(a,b') + p_{12}^{+0}(a',b) \\ & + p_{12}^{0+}(a',b) + \tfrac{1}{2}p_{12}^{00}(a',b) + p_{12}^{+0}(a',b') + p_{12}^{0+}(a',b') + \tfrac{1}{2}p_{12}^{00}(a',b') - p_1^0(a') - p_2^0(b)]\,. \end{aligned}$$

If we could establish that $Q \ge 0$, then the upper bound of (B6) would be identical to the experimentally useful upper bound of (4). No reason is immediately apparent that this is so, and unfortunately the terms p_{12}^{00}, p_1^0, and p_2^0 occurring in Q are unobservable, since they are probabilities of nothing happening. Thus auxiliary assumptions, and/or auxiliary experimentation as well, are necessary to test (B2) or, equivalently, (B6).

However, if Bell's formulation is modified *at the beginning,* his method of proof can be employed to obtain (4). Consider, instead of the correlation function (B1), the function

$$\begin{aligned} P'(a,b) &\equiv \int_\Gamma d\lambda\, \rho(\lambda)[2p_1^+(\lambda,a) - 1][2p_2^+(\lambda,b) - 1] \\ &= 4p_{12}^{++}(a,b) - 2p_1^+(a) - 2p_2^+(b) + 1\,. \end{aligned} \tag{B7}$$

Since each square bracket in P' is bounded by −1 and +1, Bell's method is applicable and yields

$$-2 \le P'(a,b) - P'(a,b') + P'(a',b) + P(a',b') \le 2. \tag{B8}$$

Using Eq. (B7), inequalities (B8) become

$$\begin{aligned} -1 \le p_{12}^{++}(a,b) - p_{12}^{++}(a,b') + p_{12}^{++}(a',b) \\ + p_{12}^{++}(a',b') - p_1^+(a') - p_2^+(b) \le 0\,, \end{aligned}$$

which, suppressing superscripts, is (4). In the context of a double-channel experiment, the three other sets of inequalities given in (B3) can be obtained in similar fashion.[29]

---

*Work by one of us (J.F.C.) was supported by the U. S. Atomic Energy Commission.

[1]J. S. Bell, Physics (N.Y.) 1, 195 (1965).

[2]K. Popper, in *Perspectives in Quantum Theory: Essays in Honor of Alfred Landé,* edited by W. Yourgrau and A. Van der Merwe (MIT Press, Cambridge, Mass., 1971), p. 182. For a reply to this conjecture, see J. S. Bell, Science 177, 880 (1972).

[3]Clauser *et al.* (Ref. 5) did relax the experimentally unattainable conditions of the *Gedankenexperiment,* but nevertheless retained determinism as a separate hypothesis.

[4]J. S. Bell, in *Foundations of Quantum Mechanics, Proceedings of the International School of Physics "Enrico*

Fermi," *Course XLIX,* edited by B. d'Espagnat (Academic, New York, 1971), p. 171. For a discussion of the experimental testability of his results, see Appendix B.

[5]J. F. Clauser, M. A. Horne, A. Shimony, and R. A. Holt, Phys. Rev. Lett. 23, 880 (1969); J. F. Clauser, Bull. Am. Phys. Soc. 14, 578 (1969); A. Shimony, in *Foundations of Quantum Mechanics* (Ref. 4), p. 182; M. A. Horne, Ph.D. thesis, Boston University, 1970 (unpublished).

[6]S. J. Freedman and J. F. Clauser, Phys. Rev. Lett. 28, 938 (1972); S. J. Freedman, Ph.D. thesis, University of California, Berkeley, 1972 [Lawrence Berkeley Laboratory Report No. LBL-391 (unpublished)]. See also Ref. 24.

[7]A. Shimony, in *Foundations of Quantum Mechanics* (Ref. 4), p. 191.

[8]We use the word "particle" in the conventional way to identify experimental phenomena exemplifying the general situation studied. Thus we do not assume here the existence of a microscopic entity with a "particle-like" structure.

[9]The practical criterion for "coincident counts" always involves a coincidence time window τ: Pairs of counts separated in time by less than τ are defined to be coincident. This procedure may appear to make the definition of $N_{12}(a,b)$ ambiguous since in general it will depend upon the experimenter's choice for τ. However, this dependence is usually insensitive to variations in τ which satisfy $s \ll \tau \ll 1/r$, where r is the average count rate at either detector, and s is the typical time separation of "true" coincidence pairs. Thus, we tacitly require the experimental arrangement to be such that this condition obtains (suitable source strength, time separation of pairs, etc.). If a sufficiently weak source is used, the ratio of "chance" coincident counts to "true" coincident counts can be made arbitrarily small, and the corresponding dead time can also be minimized.

[10]For the case where the emissions are spatially localized objects in flight from the source to each apparatus, such an intermediate time clearly exists. If the emissions are spatially extended systems (fields), which may take substantial time to exit the source, we consider the system after the emission process has commenced but before the leading edge has impinged on either apparatus.

[11]Even though we have introduced λ as the state of a specific *single* system, the assumed objectivity of the system described by this state allows us to consider an ensemble of these, physically identical to the extent that they are all characterized by the same λ. The probabilities are to be associated with this ensemble. Clearly, this procedure is conceptually sound, even in cases where we cannot in practice prepare the pure λ ensemble.

[12]Previous discussions (Refs. 1, 4, and 5) consider separately the passage of the emission through the analyzer and the detection of the emission after it has passed. Since "passage through analyzer" is not directly observable for microsystems, we consider as a unit the complete causal process from initial emission to final macroscopic effect (count). Moreover, the concept "passage through analyzer" is inappropriate for many field theories (classical, quantum-mechanical, or others) in which the emission may partially pass the analyzer.

[13]We should emphasize that an assumption is made here. By writing the density as $\rho(\lambda)$, instead of a more general conditional $\rho(\lambda|a,b)$, we deny such objective and local possibilities as these:

(a) Systems of some type originate the source which, reflecting off the analyzers and returning to the source, significantly effect the ensemble in a manner dependent upon a and b.

(b) Systems originate at the analyzers and impinge upon the source, thus effecting the ensemble in a manner dependent upon a and b.

(c) Systems originate within the intersection of the backward light cones of both analyzers and the source. These propagate into the spatial region of the whole apparatus, and simultaneously effect both the experimenters' selections of analyzer orientations and the emissions from the source.

In principle, (a) and (b) can be ruled out experimentally by rapidly and repeatedly changing the analyzer orientations immediately before each data-collection period and then stopping each data-collection period before the new orientations can be communicated (at the speed of light) to the source. The more general type-(c) conspiracies are difficult to rule out experimentally. But, for example, if the orientations a and b are selected at random by two physicists supposedly acting independently, significant cases of (c) require that their selections are *not* independent but are, in fact, strongly correlated with each other as well as with the source emissions.

[14]Because the form of λ is unspecified, we make no commitment to any specific field model, but adopt a quite general view that a field is an objective physical entity extended throughout space.

[15]Recall footnotes 10, 11, and 13 for additional informal justification of the name. The class of OLT includes the deterministic local hidden-variable theories discussed by Bell (Ref. 1) and Clauser *et al.* (Ref. 5) and is essentially the class of stochastic hidden-variable theories "with a certain local character" considered by Bell (Ref. 4). Hence we drop the name "hidden-variable theory," since traditionally the term has been used to identify theories characterized essentially by dispersion-free states, i.e., deterministic theories.

[16]Jauch [in *Foundations of Quantum Mechanics* (Ref. 4), p. 22] makes a similar conjecture when he suggests that "On the whole, and in spite of pretensions to the contrary, physicists are usually more inclined to Realism than to Positivism." However, for a specific example of implicit support for (2′), see M. L. Goldberger and K. M. Watson, Phys. Rev. 134, B919 (1964). Considering a system composed of two spin-$\frac{1}{2}$ particles, they comment: "In general therefore the observation of the orientation of spin 1 with respect to the axis has an instantaneous effect on the state of the second spin. With the interpretation of an observation as making a selection among the members of an ensemble, this is in no sense surprising." But a denial of any instantaneous effect for each member of this ensemble is tantamount to assuming (2′). The suprise, then, for an advocate of OLT, is that (2′) and quantum mechanics are incompatible.

[17]This identification, for the restricted case that each state of the mixture is a product of single-particle *quantum-mechanical* states, was conjectured by W. H. Furry [Phys. Rev. 49, 393 (1936); 49, 476 (1936)] and shown to be in disagreement with experiment by D. Bohm and Y. Aharonov [Phys. Rev. 108, 1070 (1957)]. An OLT is the generalization to mixtures of *any* type of product states.

[18]D. Bohm, *Quantum Theory* (Prentice-Hall, Englewood Cliffs, N. J., 1951), p. 614.

[19]C. S. Wu and I. Shaknov, Phys. Rev. 77, 136 (1950).

[20]L. Kasday, J. Ullman, and C. S. Wu, Bull. Am. Phys. Soc. 15, 586 (1970); see also L. Kasday, in *Foundations of Quantum Mechanics* (Ref. 4), p. 195.

[21]Actually, in atomic cascade experiments, the failure to violate (4′) is much worse than indicated in (6), since the singles probabilities p_1 and p_2 in (5) do not include the large contributions due to unpaired emissions from each transition of the cascade.

[22]The experiment of Kasday, Ullman, and Wu (Ref. 20) has confirmed the quantum-mechanical predictions for coincident Compton scattering of photon pairs produced in positronium annihilation. However, explicit OLT models exist (Refs. 5 and 20) which demonstrate that the analyzers (Compton polarimeter) are too inefficient to reveal a discrepancy between general OLT and quantum mechanics.

[23]Semiclassical radiation theories are OLT which satisfy the no-enhancement assumption. For an earlier discussion of these theories and two-photon correlations, see J. F. Clauser, Phys. Rev. A 6, 49 (1972). See also J. F. Clauser, Phys. Rev. D 9, 853 (1974).

[24]Experimental results in conflict with those of Freedman and Clauser have been found by R. A. Holt [Ph.D. thesis, Harvard University, 1973 (unpublished)]. He carried out the design of Ref. 5, as they did, but with a different atomic source of photon pairs and different optical arrangements. Holt's data agree with inequality (11′) and differ significantly from the predicted quantum-mechanical values. Further work is being pursued at various laboratories to explore this discrepancy.

[25]The upper bound on c_2, necessary to keep $p_2(\lambda, b)$ sensible, depends on the analyzer efficiencies ϵ_M^2 and ϵ_m^2 (through the functions ϵ_+^2, ϵ_-^2, and δ) and is independent of the detector efficiency η_2. On the other hand, a physical interpretation of Eq. (16) suggests that $c_2 = \eta_2$. Then if both the detector efficiencies are larger than 0.38, some probability of the model will not be sensible. However, there are undoubtedly OLT models that remain sensible for substantially larger efficiencies. Moreover, the identification $c_2 = \eta_2$ is not a logical necessity. The quantum efficiency definition for a photomultiplier depends on a long chain of experimentation and accepted theory. Recall the related comments concerning N in Sec. III.

[26]A qualitative argument with conclusions similar to ours has been given by B. d'Espagnat, in *Proceedings of the 1972 Trieste Conference on the Physicist's Conception of Nature*, edited by J. Mehra (Reidel, Dordrecht, Holland, 1973). See also B. d'Espagnat, *Conceptional Foundations of Quantum Mechanics* (Benjamin, Menlo Park, California, 1971), Chaps. 8, 9, and 16.

[27]Bell actually considers spin-$\frac{1}{2}$ pairs; we give the photon analogy.

[28]Bell does not write Eqs. (B1), but this definition is implicit in his discussion.

[29]If the no-count outcomes (denoted 0) are considered as well as the + and − ones, there are nine sets of inequalities. But only the four given in (B3) are of immediate experimental utility, since the other five all contain probabilities of nothing happening.

PHYSICAL REVIEW LETTERS

VOLUME 49 27 SEPTEMBER 1982 NUMBER 13

Bell Inequalities with a Range of Violation that Does Not Diminish as the Spin Becomes Arbitrarily Large

Anupam Garg and N. D. Mermin
Laboratory of Atomic and Solid State Physics, Cornell University, Ithaca, New York 14853
(Received 17 June 1982)

Some new, powerful, and very simple Bell inequalities are derived. They establish that for two spin-s particles in the singlet state, the quantum-theoretic Einstein-Podolsky-Rosen correlations for spin measurements along all pairs of directions from a set of N axes are incompatible with local realism for any set of distinct coplanar axes when $N=3$, and for any set of distinct axes whatever when $N=4$, with the possible exception of sets restricted by one of the constraints $\hat{a}\pm\hat{b}\pm\hat{c}\pm\hat{d}=0$. These results hold for any value of the spin.

PACS numbers: 03.65.Bz, 02.50.+s

In the spin-s generalization of Bohm's[1] spin-$\frac{1}{2}$ version of the Einstein-Podolsky-Rosen[2] experiment, the spin components m and m' of two spin-s particles in a spin singlet state φ are measured along directions $\hat{a}$ and $\hat{a}'$.[3-5] The quantum-theoretic joint distribution $q_{\hat{a}\hat{a}'}(m,m')$ vanishes whenever $\hat{a}=\hat{a}'$ unless $m'=-m$. It is thus possible to predict with certainty the result of measuring the spin of one particle along any direction, by measuring the spin of the other (which can be far away) along that direction. Einstein, Podolsky, and Rosen would account for this by an assumption of "local realism"—namely that the particles are characterized by functions μ and μ' that specify the outcome of spin measurements along any direction, and are related by

$$\mu'(\hat{a})=-\mu(\hat{a}). \tag{1}$$

Local realism is, of course, grossly incompatible with the metaphysical position of the quantum theory. In 1964 Bell[6] showed for $s=\frac{1}{2}$ that local realism is also numerically incompatible with the quantum theory, in that it requires groups of three joint distributions for three sets of correlation experiments, $q_{\hat{a}\hat{b}}$, $q_{\hat{a}\hat{c}}$, and $q_{\hat{b}\hat{c}}$, to satisfy an inequality that the quantum-theoretic distributions violate for appropriate sets of three distinct[7] axes $\hat{a}$, $\hat{b}$, and $\hat{c}$.

It was recently shown[3] that local realism implies similar inequalities for these three distributions in the general spin-s case, and that for any value of s, there are sets of coplanar directions $\hat{a}$, $\hat{b}$, $\hat{c}$, for which the spin-s inequalities are violated by the quantum-theoretic distributions. It was noted that the range of coplanar axes $\hat{a}$, $\hat{b}$, $\hat{c}$ for which the inequalities are violated shrinks to zero in the classical ($s\to\infty$) limit. Subsequently,[4] however, theoretical evidence was found that this vanishing of the range of violation with increasing s might be an artifact of the particular *ad hoc* numerical inequality used in Ref. 3 to generalize Bell's argument.

Quite recently Ögren[5] has derived several new sets of spin-s Bell inequalities for coplanar directions $\hat{a}$, $\hat{b}$, and $\hat{c}$. He finds a wider range of coplanar axes for which the quantum-theoretic distributions $q_{\hat{a}\hat{b}}$, $q_{\hat{b}\hat{c}}$, and $q_{\hat{a}\hat{c}}$ violate his inequalities, though this range also approaches zero in

VOLUME 49, NUMBER 13 PHYSICAL REVIEW LETTERS 27 SEPTEMBER 1982

the limit of infinite s, the appropriate angular measure, in one of his cases, vanishing as $s^{-1/2}$ in contrast to an s^{-1} behavior for the inequalities of Ref. 3.

We confirm below that any such vanishing of the range of violating geometries as the classical limit is approached must indeed be an artifact of the particular analytical trick employed in the argument. We show that local realism implies another quite simple set of Bell inequalities which are violated by the quantum-theoretic distributions $q_{\hat{a}\hat{b}}$, $q_{\hat{b}\hat{c}}$, and $q_{\hat{c}\hat{a}}$ for *any* set of three distinct coplanar axes, $\hat{a}$, $\hat{b}$, $\hat{c}$, *whatever the value of the spin.*

Thus for three distinct coplanar axes the extraordinary aspect of the quantum-theoretic pair distributions q associated with them holds for any choice of axes and for any value of the spin. It is known,[8] however, that the general three-axis configuration space has regions of nonvanishing volume which contain sets of noncoplanar axes for which no three-axis Bell inequality of any kind can be violated by the quantum-theoretic distributions. We shall show below that if one admits a fourth axis, the power of the inequalities is not subject to such a limitation: We show that local realism implies a family of four-axis Bell inequalities satisfied by the six distributions $q_{\hat{a}\hat{b}}$, $q_{\hat{a}\hat{c}}$, $q_{\hat{a}\hat{d}}$, $q_{\hat{b}\hat{c}}$, $q_{\hat{b}\hat{d}}$, and $q_{\hat{c}\hat{d}}$, which hold for *any* spin s, but are violated by the quantum-theoretic distributions for all four-axis sets $\hat{a}$, $\hat{b}$, $\hat{c}$, $\hat{d}$ except for a simple zero-volume subset of the four-axis configuration space.

Our results (and those of Refs. 3–6) are based on the fact that if the particles were indeed characterized by functions of direction μ and μ' specifying the subsequent results of measuring their spins along any axes, then each separate particle could be characterized by an N-axis distribution $p_{\hat{a}_1\cdots\hat{a}_N}(m_1,\ldots,m_n)$ giving the probability that the function characterizing that particle had the values m_i in the directions $\hat{a}_i$, $i=1,\ldots,n$. If such functions existed, it would follow from the relation (1) between μ and μ' that for $N=2$ they would be given (for particle 1) in terms of the quantum-theoretic distribution q characterizing the pair correlation experiment by

$$p_{\hat{a}_1\hat{a}_2}(m_1,m_2)=q_{\hat{a}_1\hat{a}_2}(m_1,-m_2). \quad (2)$$

The quantum theory, of course, emphatically denies that the distributions p_N have any meaning for a single particle, since they purport to describe the joint distribution of several distinct components of a single spin operator. The point of Bell's argument and the argument that follows is that the existence of such distributions is not only philosophically incompatible with the quantum theory, but also quantitatively incompatible with its numerical predictions. The strategy of such arguments is as follows.

From a purely statistical point of view there is nothing objectionable about the two-axis distributions $p_{\hat{a}_1\hat{a}_2}$ defined by (2); they are nonnegative, return the observed one-axis distributions as marginals, and, in conjunction with the assumption (1) of local realism, account for the observed two-particle correlations contained in $q_{\hat{a}_1\hat{a}_2}$. The difficulty, on purely statistical grounds, comes from the additional requirement that there should be N-axis distributions p with $N>2$ that return the two-axis distributions as marginals. Thus for any three axes $\hat{a}$, $\hat{b}$, $\hat{c}$, there should exist a nonnegative function $p_{\hat{a}\hat{b}\hat{c}}$ satisfying

$$p_{\hat{a}\hat{b}}(m_1,m_2)=\sum_{m_3}p_{\hat{a}\hat{b}\hat{c}}(m_1,m_2,m_3) \quad (3)$$

and the analogous two equations for the distributions $p_{\hat{a}\hat{c}}$ and $p_{\hat{b}\hat{c}}$. By a three-axis Bell inequality we mean a condition on three axes $\hat{a}$, $\hat{b}$, $\hat{c}$, which is necessary (but not, in general, sufficient[9]) for the existence of any nonnegative $p_{\hat{a}\hat{b}\hat{c}}$ satisfying the relations (3). The condition applies to the observed distributions $q_{\hat{a}\hat{b}}$, $q_{\hat{b}\hat{c}}$, and $q_{\hat{a}\hat{c}}$, since these are directly related to the hypothetical $p_{\hat{a}\hat{b}\hat{c}}$ through (2) and (3). The condition comes in the form of an inequality since it is easy to construct functions $p_{\hat{a}\hat{b}\hat{c}}$ satisfying the conditions (3); the difficulty lies in finding nonnegative functions.

The Bell inequalities given below establish for any spin s that for any distinct coplanar axes $\hat{a}$, $\hat{b}$, $\hat{c}$, the quantum-theoretic distributions $q_{\hat{a}\hat{b}}$, $q_{\hat{a}\hat{c}}$, and $q_{\hat{b}\hat{c}}$ are numerically incompatible with the existence of any nonnegative function $p_{\hat{a}\hat{b}\hat{c}}$. They also show for any spin s that a necessary condition for there to be a four-axis distribution $p_{\hat{a}\hat{b}\hat{c}\hat{d}}$ satisfying

$$p_{\hat{a}\hat{b}}(m_1,m_2)=\sum_{m_3,m_4}p_{\hat{a}\hat{b}\hat{c}\hat{d}}(m_1,m_2,m_3,m_4), \quad (4)$$

and the analogous five equations for $p_{\hat{a}\hat{c}}$, $p_{\hat{a}\hat{d}}$, $p_{\hat{b}\hat{c}}$, $p_{\hat{b}\hat{d}}$, and $p_{\hat{c}\hat{d}}$, is that the four axes satisfy one of the eight constraints $\hat{a}\pm\hat{b}\pm\hat{c}\pm\hat{d}=0$.

The only properties of the quantum-theoretic distributions needed to establish these results are the following:

(a) *Bilinearity of the correlation functions.*—The mean of the product mm',

$$\langle mm'\rangle_{\hat{a}\hat{b}}=\sum_{mm'}mm'q_{\hat{a}\hat{b}}(m,m')$$
$$=(\varphi,(\vec{S}^1\cdot\hat{a})(\vec{S}^2\cdot\hat{b})\varphi), \quad (5)$$

VOLUME 49, NUMBER 13 PHYSICAL REVIEW LETTERS 27 SEPTEMBER 1982

is explicitly bilinear in the axes $\hat{a}$ and $\hat{b}$.

(b) *Rotational invariance of the singlet state.*—This requires that the bilinear form in (5) must in fact be proportional to $\hat{a}\cdot\hat{b}$[10]:

$$\langle mm'\rangle_{\hat{a}\hat{b}} = K_s(\hat{a}\cdot\hat{b}). \tag{6}$$

(c) *One-particle distributions independent of the far detector.*—This requires that

$$\textstyle\sum_{m'} q_{\hat{a}\hat{b}}(m,m') \text{ is independent of } \hat{b}. \tag{7}$$

(d) *Perfect correlations for parallel axes.*—This is the basis for the Einstein-Podolsky-Rosen argument:

$$q_{\hat{a}.\hat{a}}(m,m') = \delta_{m,-m'}. \tag{8}$$

(e) *Positivity of the extremal distributions.*—We have

$$q_{\hat{a}.\hat{b}}(m,m') \neq 0 \text{ if } |m| \text{ or } |m'| = s, \quad \hat{a} \neq \pm\hat{b}. \tag{9}$$

This follows from the explicit form for q, which is just the binomial distribution when one of the variables is fixed at $\pm s$.

Note that of the five properties (a)–(e), only (e) fails to hold for the classical joint distribution.

The proofs of these results are surprisingly simple:

(1) *Nonexistence of distributions for three distinct coplanar axes.*—If $\hat{a}$, $\hat{b}$, and $\hat{c}$ are distinct coplanar axes then there is a relation

$$A\hat{a} + B\hat{b} + C\hat{c} = 0 \tag{10}$$

with nonzero coefficients. Label the axes so that $|A| \geq |B| \geq |C|$. Given any candidate for a three-axis distribution, $p_{\hat{a}\hat{b}\hat{c}}$, define

$$f_{\hat{a}\hat{b}\hat{c}} = \sum_{m_1 m_2 m_3} (Am_1 + Bm_2 + Cm_3)^2 p_{\hat{a}\hat{b}\hat{c}}(m_1,m_2,m_3). \tag{11}$$

Since the sum is term by term nonnegative it is bounded below by any partial sum. In particular,

$$f_{\hat{a}\hat{b}\hat{c}} \geq \textstyle\sum_{m_3}(|A|s + |B|s + Cm_3)^2 p_{\hat{a}\hat{b}\hat{c}}(\epsilon_A s, \epsilon_B s, m_3), \tag{12}$$

where $\epsilon_X = X/|X|$. If we replace the squared trinomial by its lower bound $(|A| + |B| - |C|)^2 s^2$, then the sum on m_3 gives a two-axis distribution, and we arrive at the lower bound:

$$f_{\hat{a}\hat{b}\hat{c}} \geq (|A| + |B| - |C|)^2 s^2 q_{\hat{a}\hat{b}}(\epsilon_A s, -\epsilon_B s). \tag{13}$$

Property (e) requires this lower bound to exceed zero. Explicitly squaring the trinomial in Eq. (11), however, yields nine terms, each of which can be expressed in terms of two-axis distributions, and hence in terms of q. Properties (a)–(d) permit each term to be evaluated, and the nine can then be recombined to give[11]

$$f_{\hat{a}\hat{b}\hat{c}} = -K_s(A\hat{a} + B\hat{b} + C\hat{c})^2, \tag{14}$$

which vanishes in view of (10). There can therefore be no $p_{\hat{a}\hat{b}\hat{c}}$.

(2) *Nonexistence of distributions for any four distinct axes satisfying* $\hat{a} \pm \hat{b} \pm \hat{c} \pm \hat{d} \neq 0$.—If $p_{\hat{a}\hat{b}\hat{c}\hat{d}}$ exists then so do the three-axis distributions obtained from it as marginals, and therefore by the above result no three of the axes can lie in a single plane. There therefore exists a relation

$$A\hat{a} + B\hat{b} + C\hat{c} + D\hat{d} = 0 \tag{15}$$

with nonzero coefficients. Label the axes so that $|A| \geq |B| \geq |C| \geq |D|$. Define

$$f_{\hat{a}\hat{b}\hat{c}\hat{d}} = \sum_{m_1 \cdots m_4} (Am_1 + Bm_2 + Cm_3 + Dm_4)^2 p_{\hat{a}\hat{b}\hat{c}\hat{d}}(m_1,m_2,m_3,m_4). \tag{16}$$

As in the three-axis case, we now argue that

$$f_{abcd} \geq \sum_{m_3 m_4} (|A|s + |B|s + Cm_3 + Dm_4)^2 p_{\hat{a}\hat{b}\hat{c}\hat{d}}(\epsilon_A s, \epsilon_B s, m_3 m_4) \geq (|A| + |B| - |C| - |D|)^2 s^2 q_{\hat{a}\hat{b}}(\epsilon_A s, \epsilon_B s). \tag{17}$$

In view of property (e) this last lower bound can only be zero if $|A| = |B| = |C| = |D|$. But the procedure that led to (14) now gives

$$f_{\hat{a}\hat{b}\hat{c}\hat{d}} = K_s(A\hat{a} + B\hat{b} + C\hat{c} + D\hat{d})^2, \tag{18}$$

which vanishes in view of (15). Therefore $p_{\hat{a}\hat{b}\hat{c}\hat{d}}$ can exist only when the relation (15) reduces to one of the forms $\hat{a} \pm \hat{b} \pm \hat{c} \pm \hat{d} = 0$.

Note that for spin $\frac{1}{2}$ the excluded geometries are not artifacts. The four-axis distribution

$$p_{\hat{a}\hat{b}\hat{c}\hat{d}}(m_1,m_2,m_3,m_4) = -\tfrac{1}{16} + m_1 m_2 m_3 m_4 + \tfrac{1}{8}(m_1\hat{a} + m_2\hat{b} + m_3\hat{c} + m_4\hat{d})^2 \tag{19}$$

VOLUME 49, NUMBER 13 PHYSICAL REVIEW LETTERS 27 SEPTEMBER 1982

gives all the correct two-axis distributions and is nonnegative when $\hat{a}+\hat{b}+\hat{c}+\hat{d}=0$.[12]

For $s>\frac{1}{2}$, four-axis distributions do not in general exist even in the excluded geometries. For example, let $s=1$ and let $\hat{a}$, $\hat{b}$, $\hat{c}$, and $\hat{d}$ point from the center to the vertices of a tetrahedron, so that (15) holds with $A=B=C=D=1$. In this geometry the two-axis distributions are independent of the axis pair, and are given by $p(m_1, m_2)=\frac{1}{27}$, $m_1=m_2$; $p(m_1,m_2)=\frac{4}{27}$, $m_1\neq m_2$. It suffices to consider four-axis distributions symmetric in $\hat{a}$, $\hat{b}$, $\hat{c}$, and $\hat{d}$,[13] and hence independent of the order of the four arguments m_i. The vanishing of $f_{\hat{a}\hat{b}\hat{c}\hat{d}}$ required by (18) means, in view of the definition (16), that the four-axis p must vanish unless $m_1+m_2+m_3+m_4=0$. We therefore have $\frac{1}{27}=p(1,1)=p(1,1,-1,-1)$. Furthermore, $\frac{4}{27}=p(1,-1)=p(1,-1,0,0)+2p(1,-1,1,-1)$, and so we must have $p(1,-1,0,0)=\frac{2}{27}$. But $p(0,0)=p(0,0,0,0)+2p(0,0,1,-1)$, which requires $p(0,0)$ to exceed $\frac{4}{27}$. Since the actual value is $\frac{1}{27}$, there can be no four-axis distribution.[14]

We conclude that the necessary emergence of local realism in the classical limit is not signaled by the vanishing of the range of geometries that violate Bell inequalities. For the inequalities derived here it follows, instead, from the vanishing magnitude of those violations.[15] In this regard it should be noted that there are lower bounds that vanish far more slowly with increasing s than the one given in (13). We consider this point and the generalization of our approach to the weaker form of local realism tested by the inequalities of Clauser and Horne[16] in a subsequent publication.[17]

This work was supported by the National Science Foundation under Grant No. DMR 80-20429.

---

[1]D. Bohm, *Quantum Theory* (Prentice-Hall, Englewood Cliffs, N.J., 1951), pp. 614–619.

[2]A. Einstein, B. Podolsky, and N. Rosen, Phys. Rev. 47, 777 (1935).

[3]N. D. Mermin, Phys. Rev. D 22, 356 (1980).

[4]N. D. Mermin and Gina M. Schwarz, Found. Phys. 12, 101 (1982).

[5]Mats Ögren, to be published.

[6]J. S. Bell, Physics (N.Y.) 1, 195 (1964).

[7]We say $\hat{n}$ and $\hat{n}'$ are "distinct" if $\hat{n}\neq\pm\hat{n}'$.

[8]For spin-$\frac{1}{2}$ this was shown by E. P. Wigner, Am. J. Phys. 38, 1005 (1970). For the spin-1 case see Ref. 4, Sec. 5.2.

[9]Wigner (Ref. 8) gives necessary *and* sufficient conditions for spin $\frac{1}{2}$; Mermin and Schwarz (Ref. 4, Sec. 5.2), for spin 1.

[10]It is easy to show that $K_s=-s(s+1)/3$, but this is not exploited in any of our arguments.

[11]To evaluate the three terms in which only a single m appears using only properties (a)–(d), use (7) to set the two axes equal, (8) to replace m^2 by $-mm'$, and (6) to give the result.

[12]For the other exceptional geometries simply reverse the sign of each m_i in (19) corresponding to each axis with a negative sign in $\hat{a}\pm\hat{b}\pm\hat{c}\pm\hat{d}=0$. These four-axis distributions can be shown to be the only ones possible.

[13]If an asymmetric one existed, symmetrizing it in the four axes would yield a symmetric one with all the required properties.

[14]It is shown in Ref. 4 [see Eq. (6.8)] that the three-axis spin-1 distributions do exist when the set $\hat{a}$, $\hat{b}$, $\hat{c}$ contains three of the four tetrahedral directions. For spin 1, four tetrahedral axes therefore provide an example of a geometry for which no four-axis distribution exists, even though all possible Bell inequalities must be satisfied for any three-axis subset of the four.

[15]The vanishing of the magnitude, however, may also be an artifact of the particular inequalities we have derived here, since they all follow from taking the mean of slowly varying functions of the spin components m, and are therefore insensitive to the rapidly varying parts of the quantum-theoretic distributions $q_{\hat{a}\hat{a}'}(m,m')$.

[16]J. F. Clauser and M. A. Horne, Phys. Rev. D 10, 526 (1974).

[17]Anupam Garg and N. D. Mermin, to be published.

VOLUME 49, NUMBER 17 PHYSICAL REVIEW LETTERS 25 OCTOBER 1982

ERRATUM

BELL INEQUALITIES WITH A RANGE OF VIOLATION THAT DOES NOT DIMINISH AS THE SPIN BECOMES ARBITRARILY LARGE. Anupam Garg and N. D. Mermin [Phys. Rev. Lett. 49, 901 (1982)].

An overall factor of $(2s+1)^{-1}$ should appear on the right-hand side of Eq. (8), and an overall minus sign on the right-hand side of Eq. (18); in Eq. (17) ϵ_B on the right-hand side should be preceded by a minus sign; and in Eqs. (8) and (9) there should be no comma between the subscripts of the q. Equation (7) characterizing property (c) should be supplemented by a similar equation asserting that the sum on m is independent of $\hat{a}$. These corrections are of no significance for any of the points being made. The meaning of the first sentence in Ref. 14 was somewhat distorted by the addition of the word "the" to the first line; it should be removed.

Reprinted with permission from *Dialectica,* **39**, J.S. Bell, A. Shimony, M.A. Horne, and J.F. Clauser, "An Exchange on Local Beables," pp. 85-110, ©1985. Reprinted from *Epistemological Letters:* (9) 1976; (13) 1976; (15) 1977; and (18) 1978.

An Exchange on Local Beables*

Summary

a) Bell tries to formulate more explicitly a notion of "local causality": correlations between physical events in different space-time regions should be explicable in terms of physical events in the overlap of the backward light cones. It is shown that ordinary relativistic quantum field theory is not locally causal in this sense, and cannot be embedded in a locally causal theory.

b) Clauser, Horne and Shimony criticize several steps in Bell's argument that any theory of local "beables" is incompatible with quantum mechanics. It is contended that the Clauser-Horne derivation of a Bell-type inequality circumvents his weak steps. The Clauser-Horne derivation must assume that there are no undetected correlations between choices of controllable variables in two space-like separated regions. Methodological considerations support this assumption.

c) In response to criticism by Shimony, Horne, and Clauser, Bell tries to clarify the argument of "The theory of local beables", and to defend as permissible the hypothesis of free variables.

d) Bell's reply to an earlier criticism by Shimony, Clauser, and Horne is answered. The convergence of Bell's position towards theirs is noted.

Résumé

a) Bell tente de formuler de façon plus explicite une notion de «causalité locale»: les corrélations entre événements physiques dans des régions spatio-temporelles différentes devraient être explicables à partir d'événements physiques dans l'intersection de leurs cônes de lumières passés. Il montre que la théorie quantique des champs relativiste ordinaire n'est pas localement causale dans ce sens et ne peut donc être intégrée dans une théorie localement causale.

b) Shimony, Horne et Clauser critiquent plusieurs pas dans la démonstration de Bell que toute théorie postulant des «beables» locales est incompatible avec la mécanique quantique. On prétend que la manière dont Clauser-Horne ont établi une inégalité du type Bell évite ces points faibles. La démonstration de Clauser-Horne doit supposer qu'il n'y a pas de corrélations non détectées entre les choix des variables contrôlables dans deux régions dont la séparation est du genre espace. Des considérations méthodologiques justifient ce présupposé.

c) En réponse aux critiques de Shimony, Horne et Clauser, Bell tente de clarifier le raisonnement de «The theory of local beables» et de défendre la légitimité de l'hypothèse de variables libres.

d) Réplique de Shimony à cette réponse de Bell. La convergence entre les positions des deux parties est soulignée.

Zusammenfassung

a) Bell versucht, in mehr expliziter Weise den Begriff der «lokalen Kausalität» zu bestimmen: die Korrelationen unter physikalischen Ereignissen in verschiedenen raum-zeitlichen Ge-

* Selected papers from *Epistemological Letters Hidden Variables and Quantum Uncertainty,* ed. by F. Bonsack, Institut de la Méthode, C.P. 1081, 2501 Bienne: No. 9 (1976), No. 13 (1976), No. 15 (1977), No. 18 (1978)

bieten sollten aufgrund von physikalischen Ereignissen in der Überschneidung ihrer vergangenen Lichtkegel erklärt werden. Er zeigt, dass die gewöhnliche relativistische Quantenfeldtheorie nicht in diesem Sinne lokal kausal ist und dass sie deshalb nicht in eine lokal kausale Theorie integriert werden kann.

b) Shimony, Horne und Clauser kritisieren mehrere Schritte in Bells Beweisgang, dass jede Theorie, die lokale «beables» postuliert, mit der Quantenmechanik unvereinbar sei. Es wird behauptet, dass die Art, wie Clauser-Horne eine Ungleichung des Bellschen Typs darlegen, diese Mängel vermeidet. Das Argument setzt voraus, dass es keine unaufgedeckten Korrelationen gibt zwischen der Wahl von in zwei raumartig getrennten Gebieten kontrollierbaren Variablen. Methodologische Überlegungen rechtfertigen diese Voraussetzung.

c) Bell antwortet auf diese Kritik, indem er den Gedankengang der früheren Arbeit klärt und die Legitimität seiner Hypothese verteidigt.

d) Shimony antwortet und unterstreicht die Konvergenz der Positionen.

J.S. Bell - *The Theory of Local Beables* **

Introduction - The Theory of Local Beables

This is a pretentious name for a theory which hardly exists otherwise, but which ought to exist. The name is deliberately modelled on "the algebra of local observables". The terminology, *be*-able as against *observ*-able, is not designed to frighten with metaphysic those dedicated to realphysic. It is chosen rather to help in making explicit some notions already implicit in, and basic to, ordinary quantum theory. For, in the words of Bohr [1], "it is decisive to recognize that, however far the phenomena transcend the scope of classical physical explanation, the account of all evidence must be expressed in classical terms". It is the ambition of the theory of local beables to bring these "classical terms" into the mathematics, and not relegate them entirely to the surrounding talk.

The concept of "observable" lends itself to very precise *mathematics* when identified with "self-adjoint operator". But physically, it is a rather woolly concept. It is not easy to identify precisely which physical processes are to be given the status of "observations" and which are to be relegated to the limbo between one observation and another. So it could be hoped that some increase in precision might be possible by concentration on the *be*ables, which can be described in "classical terms", because they are there. The beables must include the settings of switches and knobs on experimental equipment, the currents in coils, and the readings of instruments. "Observables" must be *made,* somehow, out of beables. The theory of local beables should contain, and give precise physical meaning to, the algebra of local observables.

** Presented at the sixth GIFT Seminar Jaca, June 2—7, 1975

The word "beable" will also be used here to carry another distinction, that familiar already in classical theory between "physical" and "non-physical" quantities. In Maxwell's electromagnetic theory, for example, the fields $\vec{E}$ and $\vec{H}$ are "physical" (beables, we will say) but the potentials $\vec{A}$ and Φ are "non-physical". Because of gauge invariance the same physical situation can be described by very different potentials. It does not matter that in Coulomb gauge the scalar potential propagates with infinite velocity. It is not really supposed to *be* there. It is just a mathematical convenience.

One of the apparent non-localities of quantum mechanics is the instantaneous, over all space, "collapse of the wave function" on "measurement". But this does not bother us if we do not grant beable status to the wave function. We can regard it simply as a convenient but inessential mathematical device for formulating correlations between experimental procedures and experimental results, i.e., between one set of beables and another. Then its odd behaviour is as acceptable as the funny behaviour of the scalar potential of Maxwell's theory in Coulomb gauge.

We will be particularly concerned with *local* beables, those which (unlike for example the total energy) can be assigned to some bounded space time region. For example, in Maxwell's theory the beables local to a given region are just the fields $\vec{E}$ and $\vec{H}$, in that region, and all functionals thereof. It is in terms of local beables that we can hope to formulate some notion of local causality. Of course we may be obliged to develop theories in which there *are* no strictly local beables. That possibility will not be considered here.

1) *Local determinism*

In Maxwell's theory, the fields in any spacetime region 1 are determined by those in any space region V, at some time t, which fully closes the backward light cone of 1:

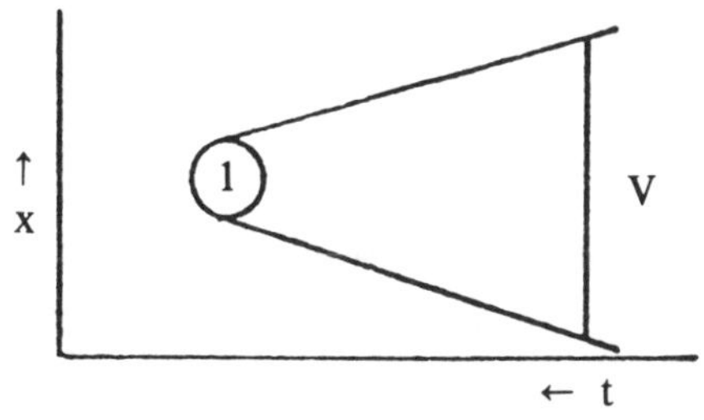

Because the region V is limited, localized, we will say the theory exhibits *local determinism*. We would like to form some notion of *local causality* in theories which are not deterministic, in which the correlations prescribed by the theory, for the beables, are weaker.

2) *Local causality*

Consider a theory in which the assignment of values to some beables Λ implies, not necessarily a particular value, but a probability distribution, for another beable A. Let

$$\{A \mid \Lambda\}$$

denote the probability of a particular value A given particular values Λ. Let A be localized in a space-time region 1. Let B be a second beable localized in a second region 2 separated from 1 in a spacelike way:

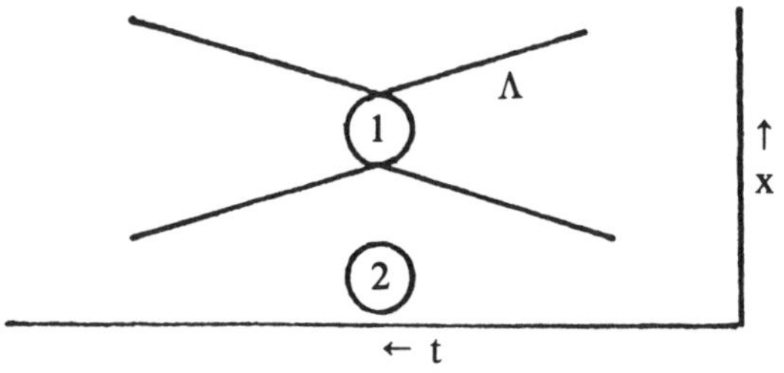

Now my intuitive notion of local causality is that events in 2 should not be "causes" of events in 1, and vice versa. But this does not mean that the two sets of events should be uncorrelated, for they could have common causes in the overlap of their backward light cones. It is perfectly intelligible then that if Λ in (1) does not contain a complete record of events in that overlap, it can be usefully supplemented by information from region 2. So in general it is expected that

$$\{A \mid \Lambda, B\} \neq \{A \mid \Lambda\} \tag{1}$$

However, in the particular case that Λ contains already a *complete* specification of beables in the overlap of the two light cones, supplementary information from region 2 could reasonably be expected to be redundant. So, with some change of notation, we formulate local causality as follows:

Let N denote a specification of *all* the beables, of some theory, belonging to the overlap of the backward light cones of spacelike separated regions 1 and 2. Let Λ be a specification of some beables from the remainder of the backward light cone of 1, and B of some beables in the region 2. Then in a *locally causal theory*

$$\{A \mid \Lambda, N, B\} = \{A \mid \Lambda, N\} \tag{2}$$

whenever both probabilities are given by the theory.

3) *Quantum mechanics is not locally causal*

Ordinary quantum mechanics, even the relativistic quantum field theory, is not locally causal in the sense of (2). Suppose, for example, we have a radioactive nucleus which can emit a single α-particle, surrounded at a considerable distance by α-particle counters. So long as it is not specified that some *other* counter registers, there is a chance for a particular counter that *it* registers. But if it is specified that some other counter does register, even in a region of space-time outside the relevant backward light cone, the chance that the given counter registers is zero. We simply do not have (2). Could it be that here we have an incomplete specification of the beables N? Not so long as we stick to the list of beables recognized in ordinary quantum mechanics — the settings of switches and knobs and currents needed to prepare the initial unstable nucleus. For these are completely summarized, in so far as they are relevant for predictions about counter registering, in so far as such predictions are possible in quantum mechanics, by the wave function.

But could it not be that quantum mechanics is a fragment of a more complete theory, in which there are other ways of using the given beables, or in which there are additional beables — hitherto "hidden" beables? And could it not be that this more complete theory has local causality? Quantum mechanical predictions would then apply not to given values of all the beables, but to some probability distribution over them, in which the beables recognized as relevant by quantum mechanics are held fixed. We will investigate this question, and answer it in the negative.

4) *Locality inequality*

Consider a pair of beables A and B, belonging respectively to regions 1 and 2 with spacelike separation, which happen by definition to have the property

$$|A| \leq 1 \qquad\qquad |B| \leq 1 \tag{3}$$

Consider the situation in which beables Λ , M, N are specified, where N is a *complete* specification of the beables in the overlap of the light cones, and Λ and M belong respectively to the remainders of the two light cones

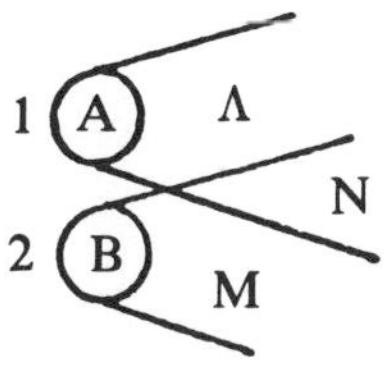

Consider the joint probability distribution

$$\{A, B \mid \Lambda, M, N\} \tag{4}$$

By a standard rule of probability, it is equal to

$$\{A \mid \Lambda, M, N, B\}\{B \mid \Lambda, M, N\} \tag{5}$$

which, by (2), is the same as

$$\{A \mid \Lambda, N\}\{B \mid M, N\} \tag{6}$$

This says simply that correlations between A and B can arise only because of common causes N.

Consider now the expectation value of the product AB

$$p(\Lambda, M, N) = \sum_{A, B} A B \{A \mid \Lambda, N\}\{B \mid M, N\} \tag{7}$$

(where the summation stands also, if necessary, for integration)

$$= \bar{A}(\Lambda, N)\bar{B}(M, N) \tag{8}$$

where $\bar{A}$ and $\bar{B}$ are functions of the variables indicated, and

$$|\bar{A}| \leq 1 \qquad\qquad |\bar{B}| \leq 1 \tag{9}$$

for all values of the arguments. Let Λ' and M' be alternative specifications, of the same regions, to ∧ and M.

$$p(\Lambda, M, N) \pm p(\Lambda, M', N) = \bar{A}(\Lambda, N)[\bar{B}(M, N) \pm \bar{B}(M', N)] \tag{10}$$

$$p(\Lambda', M, N) \pm p(\Lambda', M', N) = \bar{A}(\Lambda', N)[\bar{B}(M, N) \pm \bar{B}(M', N)]$$

whence, using (9),

$$|p(\Lambda, M, N) \pm p(\Lambda, M', N)| \leq |\bar{B}(M, N) \pm \bar{B}(M', N)|$$
$$|p(\Lambda', M, N) \pm p(\Lambda', M', N)| \leq |\bar{B}(M, N) \pm \bar{B}(M', N)| \tag{11}$$

so that finally, again invoking (9), and $|a + b| + |a - b| \leq 2\,\mathrm{Max}(|a|, |b|)$,

$$|p(\Lambda, M, N) \pm p(\Lambda, M', N)| + |p(\Lambda', M, N) \mp p(\Lambda', M', N)| \leq 2 \tag{12}$$

Suppose now the specifications Λ, M, N are each given in two parts

$$\Lambda \equiv (a, \lambda)$$
$$M \equiv (b, \mu)$$
$$N \equiv (c, \nu)$$

where we are particularly interested in the dependence on a, b, c, while λ, μ, ν, are averaged over some probability distributions — which may depend on a, b, c. In the comparison with quantum mechanics, we will think of a, b, c, as variables which specify the experimental set-up in the sense of quantum mechanics, while λ, μ, ν, are in that sense either hidden or irrelevant.
Define

$$P(a, b, c) = \overline{p\big((a, \lambda), (b, \mu), (c, \nu)\big)} \tag{13}$$

where the bar denotes the averaging over (λ, μ, ν) just described. Now applying again the locality hypothesis (3), the distribution of λ and ν must be independent of b, μ — the latter being outside the relevant backward light cones.
So

$$| P(a, b, c) \pm P(a, b', C)| \leq \overline{| p\big((a,\lambda),(b,\mu),(c,\nu)\big) \pm p\big((a,\lambda),(b',\mu'),(c,\nu)\big)|} \tag{14}$$

- because the mod of the average is less than the average of the mod. In the same way

$$| P(a', b, c) \mp P(a', b', c)| \leq \overline{| p\big((a',\lambda'),(b,\mu),(c,\nu)\big) \mp p\big((a',\lambda'),(b',\mu'),(c,\nu)\big)|} \tag{15}$$

Finally then, from (14), (15) and (12),

$$|P(a, b, c) \mp P(a, b', c)| + | P(a', b, c) \pm P(a', b', c)| \leq 2 \tag{16}$$

5) *Quantum mechanics*

Quantum mechanics, however, gives certain correlations which *do not satisfy* the locality inequality (16).

Suppose, for example, a neutral pion is produced, by some experimental device, in some small space-time region 3. It quickly decays into a pair of photons. Suppose we have photon counters in space-time regions 1 and 2 so located with respect to 3 that when one photon falls on 1, the second falls (or nearly always does) on 2. If the π° is at rest the counters must be equally far away in opposite directions and their sensitive times appropriately delayed. Of course, both photons will often miss both counters. Suppose finally that both counters are behind filters which pass only photons with specified linear polarization, say at angles θ and ϕ respectively to some plane containing the axis joining the two counters.

Let us calculate according to quantum mechanics the probability of the various possible responses of the counters. If $|\ \theta\ \rangle$ denotes a photon linearly polarized at un angle θ, then for the photons going towards the counters the combined spin state is

$$|\,s\rangle = \frac{1}{\sqrt{2}}|0\rangle\,|\pi/2\rangle - \frac{1}{\sqrt{2}}|\pi/2\rangle\,|0\rangle \qquad (17)$$

where first and second kets in each term refer to the photons going towards regions 1 and 2, respectively. This form is dictated by considerations of parity and angular momentum. The probability that such photons pass the filters is then proportional to

$$\begin{aligned} &½\,|\,\langle\theta\,|0\rangle\,\langle\phi\,|\pi/2\rangle - \langle\theta\,|\pi/2\rangle\,\langle\phi|0\rangle\,|^2 \\ &= ½\,|\cos\theta\sin\phi - \sin\theta\cos\phi\,|^2 \\ &= ½\,|\sin(\theta\text{-}\phi)|^2 \end{aligned} \qquad (18)$$

The corresponding factor for photon 1 to pass and photon 2 not is

$$\begin{aligned} &½\,|\,\langle\theta\,|0\rangle\,\langle\phi + \pi/2|\pi/2\,\rangle - \langle\theta\,|\pi/2\rangle\,\langle\phi + \pi/2\,|\,0\rangle\,|^2 \\ &= ½\,|\cos(\theta - \phi)|^2 \end{aligned} \qquad (19)$$

and so on. The probabilities for the various possible counting configurations are then

$$\begin{aligned} \varrho\,(\text{yes, yes}) &= \frac{x\Omega}{4\pi}\,½\,|\sin(\theta-\phi)|^2 \\ \varrho\,(\text{yes, no}) &= \frac{x\Omega}{4\pi}\,½\,|\cos(\theta-\phi)|^2 \\ \varrho\,(\text{no, yes}) &= \frac{x\Omega}{4\pi}\,½\,|\cos(\theta-\phi)|^2 \\ \varrho\,(\text{no, no}) &= \frac{x\Omega}{4\pi}\,½\,|\sin(\theta-\phi)|^2 + x\,(1 - \frac{\Omega}{4\pi}) + (1-x) \end{aligned} \qquad (20)$$

where x is the probability that the π° production mechanism actually works, Ω the (small) solid angle subtended by each counter at the production point, and no allowance has been made for bad timing, bad placing, or inefficient counting.

Now let us count A = ± 1 for (yes/no) at 1 and B = ± 1 for (yes/no) at 2. Then the quantum mechanical mean value of the product is

$$\begin{aligned} P\,(\theta, \phi) &= \varrho\,(\text{yes, yes}) + \varrho\,(\text{no, no}) - \varrho\,(\text{yes, no}) - \varrho\,(\text{no, yes}) \\ &= 1 - \frac{x\Omega}{4\pi}\,(1 + \cos 2\,(\theta\text{-}\phi)) \end{aligned} \qquad (21)$$

so that

$$|P(\theta, \phi) - P\,(\theta, \phi')| + P\,(\theta', \phi) + P\,(\theta', \phi') - 2 =$$

$$\frac{x\Omega}{4\pi}\{|\cos 2\,(\theta - \phi) - \cos 2\,(\theta - \phi')| - \cos 2\,(\theta' - \phi) - \cos 2\,(\theta' - \phi') - 2\} \qquad (22)$$

The right-hand side of this expression is sometimes positive. Take in particular

$$\phi = 0, 2\theta = \frac{\pi}{4}, -2\phi' = \frac{\pi}{2}, 2\theta' = \frac{3\pi}{4} \tag{23}$$

in which case the factur in curly brackets is

$$\{ \quad \} = \frac{1}{\sqrt{2}} + \frac{1}{\sqrt{2}} + \frac{1}{\sqrt{2}} + \frac{1}{\sqrt{2}} - 2 = +2(\sqrt{2} - 1) \tag{24}$$

But if quantum mechanics were embeddable in a locally causal theory (16) would apply, with a $\rightarrow \theta$, b $\rightarrow \phi$, and c the implicit specification of the production mechanism, held fixed in (22). The right-hand side of (22) should then be *negative*. So quantum mechanics is *not* embeddable in a locally causal theory as formulated above.

6) *Experiments*

These considerations have inspired a number of experiments. The accuracy of quantum mechanics on the atomic scale makes it hard to believe that it could be seriously wrong on that scale in some hitherto undiscovered way. The ground state of the helium atom, for example, is just the kind of correlated wave function which is embarrassing, and its energy comes out right to very high accuracy. But perhaps it is sensible to verify that theses curious correlations persist over macroscopic distances.

Experiments so far performed do not at all approach the ideal in which the settings of the instruments are determined only while the particles are in flight. When they are decided in advance, in space time regions projecting into the overlap of the backward light cones, (16) does not follow from (12). For it was supposed in (12) that the complete specification N of the overlap is the same for the various cases compared. So one can imagine a theory which is locally causal in our sense but still manages to agree with quantum mechanics for static instruments. But it would have to contain a very clever mechanism by which the result registered by one instrument depends, after a suitable time lapse, on the setting of an arbitrarily distant instrument. So static experiments are also quite interesting.

Practical experiments are far removed from the ideal in other directions also. Geometrical and other inefficiencies lead to counters registering (no, no) with overwhelming probability, (yes, yes) very seldom, and (yes, no) and (no, yes) with probabilities only weakly dependent on the settings of the instruments. Then from (21)

$$P = 1 - \epsilon^2$$

with ϵ^2 weakly dependent on the variables, so that (16) is trivially satisfied. The authors in general make some more or less ad hoc extrapolation to connect the results of the pratical with the result of the ideal experiment. It is in this sense that the entirely unauthorized "Bell's limit" sometimes plotted along with experimental points has to be understood. But such experiments also are of very high interest. For if quantum mechanics is to fail somewhere, and in the absence of a monstrous conspiracy, this should show up at some point on this side of the ideal gedanken experiment.

Several of these experiments [26] [27] [28] show impressive agreement with quantum mechanics, and exclude deviations as large as might be suggested by the locality inequality. Another experiment [29], very similar to one of those quoted [26], is said to be in agreement with it and yet in dramatic disagreement with quantum mechanics! And another experiment [30] disagrees significantly with the quantum prediction. Of course any such disagreement, if confirmed, is of the utmost importance, and that independently of the kind of consideration we have been making here.

7) *Messages*

Suppose that we are finally obliged to accept the existence of these correlations at long range, and the gross non-locality of nature in the sense of this analysis. Can *we* then signal faster than light? To answer this we need at least a schematic theory of what *we* can do, a fragment of a theory of human beings. Suppose we can control variables like a and b above, but not those like A and B. I do not quite know what "like" means here, but suppose that beables somehow fall into two classes, "controllables" and "uncontrollables". The latter are no use for *sending* signals, but can be used for *reception*. Suppose that to A corresponds a quantum mechanical "observable", an operator $\mathcal{A}$. Then if

$$\delta\mathcal{A} / \delta b \neq o$$

we could signal between the corresponding space time regions, using a change in b to induce a change in the expectation value of $\mathcal{A}$ or of some function of $\mathcal{A}$.

Suppose next that what we do when we change b is to change the quantum mechanical Hamiltonian $\mathcal{H}$ (say by changing some extermal field), so that

$$\delta \int dt\, \mathcal{H} = \mathcal{B}\, \delta b$$

where $\mathcal{B}$ is again an "observable" (i.e., an operator) localized in the region 2 of b. Then it is an exercise [31] in quantum mechanics to show that if in a given reference system region (2) is entirely later in time than region (1)

$$\delta\mathcal{A} / \delta b = 0$$

while if the reverse is true

$$\delta\mathcal{A} / \delta b = [\mathcal{A}, -(^{i}/\hbar)\, \mathcal{B}]$$

which is again zero (for spacelike separation) in quantum field theory by the usual local commutativity condition.

So if the ordinary quantum field theory is embedded in this way in a theory of beables, it implies that faster than light signalling is not possible. In this *human* sense relativistic quantum mechanics *is* locally causal.

8) *Reservations and acknowledgements*

Of course the assumptions leading to (16) can be challenged. Equation (22) may not embody *your* idea of local causality. You may feel that only the "human" version of the last section is sensible and may see some way to make it more precise.

The space time structure has been taken as given here. How then about gravitation?

It has been assumed that the settings of instruments are in some sense free variables — say at the whim of experimenters — or in any case not determined in the overlap of the backward light cones. Indeed without such freedom I would not know how to formulate *any* idea of local causality, even the modest human one.

This paper has been an attempt to be rather explicit and general about the notion of locality, along lines only hinted at in previous publications [Refs. 2), 4), 10), 19)]. As regards the literature on the subject, I am particularly conscious of having profited from the paper of Clauser, Horne, Holt and Shimony[3], which gave the prototype of (16), and from that of Clauser and Horne[16]. As well as a general analysis of the topic this last paper contains a valuable discussion of how best to apply the inequality in practice; I am indebted to it in particular for the point that in two-body decays (as compared with three-) the basic geometrical inefficiencies enter in (22) in a relatively harmless way. I have also profited from many discussions of the whole subject with Professor B. d'Espagnat.

J.S. Bell, CERN — Genève

REFERENCES

1) N. Bohr, in Albert Einstein, Ed. Schilpp, Tudor (1951).
2) J.S. Bell, Physics *1,* 195 (1965).
3) J.F. Clauser, R.A. Holt, M.A. Horne and A. Shimony, Phys. Rev. Letters *23,* 880 (1969).
4) J.S. Bell, in Proceedings of the International School of Physics "Enrico Fermi", Course IL, Varenna 1970 (Academic Press 1971).
5) R. Friedberg (1969, unpublished) referred to by M. Jammer 17).
6) E.P. Wigner, Am. J. Phys. *38,* 1005 (1970).
7) B. d'Espagnat, Conceptual Foundations of Quantum Mechanics, Benjamin (1971).
8) K. Popper, in Perspectives in Quantum Theory, Eds. W. Yourgrau and A. Van der Merwe, M.I.T. Press (1971).

9) H.P. Stapp, Phys. Rev. *D3,* 1303 (1971).
10) J.S. Bell, Science *177,* 880 (1972).
11) P.M. Pearle, Phys. Rev. *D2,* 1418 (1970).
12) J.H. McGuire and E.S. Fry, Phys. Rev. *D7,* 555 (1972)
13) S. Freedman and E.P. Wigner, Foundations of Physics *3,* 457 (1973).
14) F.J. Belinfante, A Survey of Hidden Variable Theories, Pergamon (1973).
15) V. Capasso, D. Fortunato and F. Selleri, Int. J. of Theor. Phys. *7,* 319 (1973).
16) J.F. Clauser and M.A. Horne, Phys. Rev. *D10,* 526 (1974).
17) M. Jammer, The Philosophy of Quantum Mechanics, Wiley (1974). See in particular references to T.D. Lee (p. 308) and R. Friedberg (pp. 244, 309, 324).
18) D. Gutkowski and G. Masotto, Nuovo Cimento *22B,* 1921 (1974).
19) J.S. Bell, in The Physicist's Conception of Nature, Ed. J. Mehra and D. Reidel (1973).
20) B. d'Espagnat, Phys. Rev. *D11,* 1424 (1975).
21) G. Corleo, D. Gutkowski and G. Masotto, Palermo preprint (1974).
22) H.P. Stapp, Berkeley preprint (1975), LBL 3685.
23) D. Bohm and B. Hiley, Birkbeck preprint (1975).
24) A. Baracca, Firenze preprint (1975).
25) A. Baracca, D.J. Bohm, R.J. Hiley and A.E.G. Stuart, Birkbeck preprint (1975).
26) L.R. Kasday, in Proceedings of the International School of Physics "Enrico Fermi", Course IL, Varenna 1970 (Academic Press 1971); L.R. Kasday, J.D. Ullman and C.S. Wu, Columbia preprint (1975).
27) S.J. Freedman and J.F. Clauser, Phys. Rev. Letters *28,* 938 (1972).
28) M. Lamehi-Rachti and W. Mittig, Saclay preprint, Colloque sur un "Demi-siècle de la Mécanique Quantique", 2-5 May, 1974.
29) G. Faraci, D. Gutkowski, S. Notarrigo and A.R. Pennisi, Lettere al Nuovo Cimento *9,* 607 (1974).
30) R. A. Holt and F.M. Pipkin, Harvard preprint (1974).
31) R.E. Peierls, Proc. Roy. Soc. *A214,* 143 (1952).

A. Shimony, M.A. Horne, J.F. Clauser - Comment on "The Theory of Local Beables"*

Dr. Bell's paper, "The Theory of Local Beables", performs a valuable service in clarifying two fundamental concepts: namely, *locality* and *physical reality*. His clarification leads him to a fundamental and highly reasonable assumption, expressed in equation (2) of Sect. 2. He then attempts in Sect. 4 to prove inequality (16) as a consequence of his equation (2). Unfortunately, we believe that his proof is not correct. A counter-example shows that (16) does not follow from (2) alone. Our objections are not given in a spirit of skepticism, since (16) does follow from other reasonable assumptions of locality and physical reality. These assumptions were discussed in an earlier paper[1] and will be reconsidered in this letter.

To illustrate the falsity of his claim we consider the following local beable situation. A person concocts a set of correlation experiment data. The data consist of four columns of numbers, indexed by event number j. Two of the columns contain the apparatus parameter settings, a_j and b_j, while the other two columns contain the experimental results, A_j and B_j. These data have been so contrived as to exhibit the correlation specified by quantum mechanics. The person sends the result columns (A_j and B_j) to an apparatus manufacturer; he sends the apparatus parameter settings to the secretaries of two physicists who will perform a correlation experiment using apparatus supplied by the manufacturer. The manufacturer preprograms the apparatus simply to display in sequence the results A_j (B_j) independently of what parameter setting is employed by physicist 1 (2). As physicist 1 (2) is about to record the result of the jth event, his secretary quietly whispers in his ear the suggestion that he set his apparatus parameter to the value a_j (b_j). The experimentalists thus record preprogrammed results and parameter settings *which are* consistent with the quantum mechanical prediction. Thus, when they later compare their data, they find the resulting correlation is in violation of (16). Clearly, the violation occurs even though local beables alone were responsible for the results.

Now let us examine Bell's argument in some detail, in order to see what has gone amiss. We shall first recapitulate some of his notation. Recall that he is concerned with two beables A and B, localized respectively in space-time regions 1 and 2 which are space-like separated from each other. Denote by N the full set of beables contained in the region formed by the intersection of the backward light cones of 1 and 2. Denote by Λ the full set of beables in the

* Work done under the auspices of the U.S. Energy
[1] Research & Development contract No. W-7405-Eng-48.

remainder of the backward light cone of 1 and by M those in the remainder of the backward light cone of 2. (See Fig. 1).

Bell uses the notation $\{C \mid D\}$ to mean the probability (or probability density, in case of a continuum of values for C) given D of the value C for some variable beable. (He denotes the variable by same letter C, an ambiguity in notation which causes no confusion).

Bell's formulation of local causality, Eq. (2), is essentially the following: Let C be a variable beable localized in some space-time region, large or small. This region has a unique backward light cone; let D denote *all* the beables in this backward light cone. Then

$$\{C \mid D, E\} = \{C \mid D\}$$

holds for any beables E localized in space-time regions with a space-like separation from the region of C. We fully accept that this formulation is an unambiguous and concise statement of reasonable hypotheses about locality and reality. Furthermore, some of Bell's applications of Eq. (2) are certainly legitimate, specifically the replacements

$$\{A \mid \Lambda, M, N, B\} = \{A \mid \Lambda, N\}$$
$$\{B \mid \Lambda, M, N\} = \{B \mid M, N\}$$

when he proceeds from Eq. (5) to Eq. (6). We believe, however, that Bell is incorrect in drawing certain other consequences from Eq. (2).

He proceeds by dividing Λ into two parts a and λ; M into two parts b and μ; and N into two parts c and ν, where a,b,c, are controllable variables characterizing the experimental set-up and λ, μ, ν are the respective residual parts. He then speaks of probability distributions of λ, μ, ν and says that there may depend upon a, b, c, but unfortunately he does not explicitly state what depends on what, except for the remark: "Now applying again the locality hypothesis. . ., the distribution of λ and μ must be independent of b, — the latter being outside the relevant light cones". Let us now guess the dependences which Bell has in mind (subject, of course, to correction by him). By a standard rule of probability, we have

$$\{\lambda, \mu, \nu \mid a, b, c\}$$
$$= \{\lambda \mid \mu, \nu, a, b, c\} \{\mu \mid \nu, a, b, c\} \{\nu \mid a, b, c\}.$$

We conjecture that Bell now wishes to make two separate appeals to Equation (2) to obtain the following replacements;

$$\text{(i)}\ \{\lambda \mid \mu, \nu, a, b, c\} = \{\lambda \mid \nu, a, c\}$$
$$\text{(ii)}\ \{\mu \mid \nu, a, b, c\} = \{\mu \mid \nu, b, c\}$$

Both (i) and (ii) are consequences of Eq. (2) as we have reformulated it (more precisely, of slight extensions of it). For, even though the space-time region in which λ is located extends to negative infinity in time, ν, a, c are *all* the beables other than λ itself in the region of λ and in the backward light cone of this region, and μ and b *do* refer to beables with space-like separation from the λ region. Similar reasoning holds for assertion (ii).

Our guessing, however, is not finished. Bell's derivation of (16) apparently also requires a suitable assertion concerning the distribution of ν. Indeed, we see no way of filling out the outline of his argument[2] without using the following:

$$\text{(iii)}\quad \{\nu \mid \mathrm{a, b, c}\} = \{\nu \mid \mathrm{c}\}.$$

It seems obvious that (iii) does not follow from Eq. (2), since the space-time regions containing a and b do not have space-like separation from the region of ν. In fact, the forward light cone of the region containing ν fills all of space-time. Could (iii) perhaps be a reasonable extension of Eq. (2)? We think not, at least not at the extreme level of generality that Bell seeks, since ν is the complete specification of the region to which it refers (minus the one factor c) and consequently a specific value for ν could hardly fail to influence the subsequent values of a and b. As a result of such influences, the probability distribution over the phase space of ν values would in general be conditional upon the values of a and b. This dependence cannot be excluded without further argument. It seems to us that (iii) could be made reasonable only if the settings of a and b are the results of some spontaneous events, such as acts of free will of the experimenters. (As Bell may have assumed tacitly in his derivation of (16) and explicitly in Sect. 8). This is a logical and metaphysical possibility, which we do not intend to exclude a priori. But since Bell's argument is intended to be general, it would not be legitimate for him to justify the assertion (iii) by relying upon a metaphysics which has not been proved and which may well be false[3].

It should be noted that in the second paragraph of Sect. 6 of his letter, Bell expresses certain reservations about the decisiveness of experiments based on inequality (16). He emphasizes that, "it was supposed in (12) that the complete specification N of the overlap is the same for the various cases compared". These reservations are very close in spirit to the reservations which we have just now expressed against Bell's derivation of inequality (16) itself; thus, in a way he has anticipated our criticism.

We do not regard the flaws in Bell's argument as fatal to the enterprise of deriving an inequality which is valid for a reasonable class of local theories. We feel that such a derivation was given in the paper by Clauser and Horne[1].

In that paper two spatially separated analyzer-detector assemblies were considered along with a source of emissions located midway between the assemblies, with each emission presumably consisting of two objectively real components. It is to be noted that the set-up described is considerably more specific than that described by Bell, and therefore assumptions concerning it can be less general and more plausible than those needed by Bell. Let K denote the complete state of one of the emissions (denoted by λ in Ref. 1). Let p_1 (K, a) be the probability that assembly 1 registers a detection event, when K is the emission state and when an adjustable parameter of the assembly is chosen to be a. Let p_2 (K, b) be similarly defined for assembly 2. Finally, let p_{12} (K, a, b) be the probability of a joint detection event given K, a, and b. Clauser and Horne make two suppositions:[4]

(1) $p_{12}(K, a, b) = p_1(K, a)\, p_2(K, b)$,

which they justify on grounds of locality and reality similar to Bell's.

(2) The distribution ϱ (K) of the emissions is independent of the settings a and b.

Supposition (2) of Clauser and Horne plays a role in their argument analogous to Bell's assertion (iii). The central question is whether the supposition (2) is more reasonable than (iii). Our contention is that it is, though we do not pretend to offer a definitive proof nor do we think that one can be given.

It is obvious to begin with, that the assumption of Clauser and Horne is very much weaker than Bell's. In his notation, their assuption is

(2) $\{K \mid a, b, c\} = \{K \mid c\}$, which can be written out as

(2') $\int\{K \mid \nu\}\{\nu \mid a, b, c\}\, d\nu = \int \{K \mid \nu\}\{\nu \mid c\}\, d\nu$.

But this is just an integrated form of assertion (iii), $\{\nu \mid a, b, c\} = \{\nu \mid c\}$, with $\{K \mid \nu\}$ used as a weighting function in the integration. Of course, the assertion of the equality of two integrals is a much weaker statement than the equality of two integrands. But there is yet more to be said. It is well known in the statistical mechanics of extended systems that the normal dynamics of the system as well as external perturbations tend to wash out correlations between variables which are temporally or spatially well-separated, unless there are specific mechanisms of the system for maintaining these correlations. In the present case, a, b, and K are values associated with well separated events. Moreover, there is no mechanism that one can point to which sets up a correlation between the selection of parameter a (or of b) and the occurrence of an emission having state K. Therefore, even though the left hand side of (2 ') contains a factor $\{K \mid \nu\}$ and another factor $\{\nu \mid a, b, c\}$, it is reasonable that the way in which the distribution of ν is influenced by a and b is irrelevant for the distribution of K.

Bell can, of course, reply that we do not know that the distribution of emissions K is insensitive to the values of a and b, or for that matter that there are no causal links between the act of selecting a and that of selecting b. After all, the backward light cones of those two acts do eventually overlap, and one can imagine one region which controls the decision of the two experimenters who chose a and b. We cannot deny such a possibility. But we feel that it is wrong on methodological grounds to worry seriously about it if no specific causal linkage is proposed. In any scientific experiment in which two or more variables are supposed to be randomly selected, one can always conjecture that some factor in the overlap of the backward light cones has controlled the presumably random choices. But, we maintain, skepticism of this sort will essentially dismiss all results of scientific experimentation. Unless we proceed under the assumption that hidden conspiracies of this sort do not occur, we have abondoned in advance the whole enterprise of discovering the laws of nature by experimentation[5].

To sum up: the advantage of the Clauser-Horne approach over that of Bell's is not that it is supposition free. Rather, it is that the supposition needed is no stronger than one needs for experimental reasoning generically, and nevertheless just strong enough to yield the desired inequality.

A. Shimony, Dept of Physics & Philosophy,
Boston University

M.A. Horne, Dept of Physics,
Stonehill College,
North Easton

J.F. Clauser, Magnetic Fusion Devision,
L-386, Lawrence Livermore
Laboratory, Livermore

Fig. 1 Space-time diagram of correlation experiment beables.

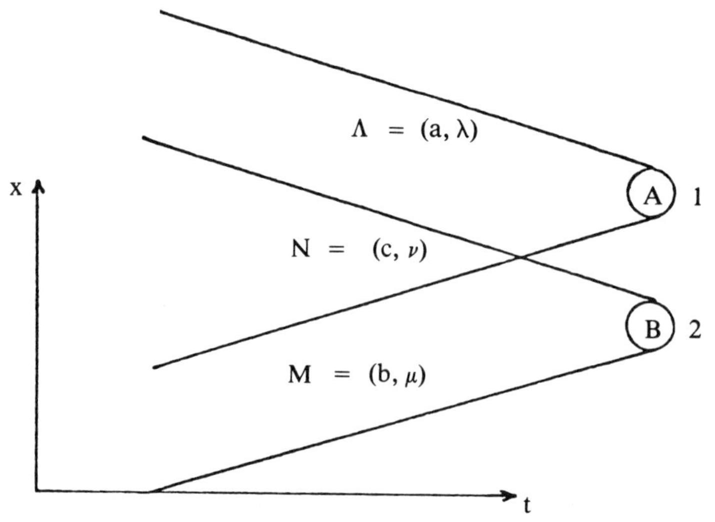

REFERENCES

1. J.F. Clauser and M.A. Horne, Phys. Rev., *D10*, 526 (1974).
2. Even given (i), (ii), and (iii) there is a slip in Bell's argument for inequality (14) is incorrect, since the distribution for μ given b is different from the distribution for μ' given b'. However, given (i), (ii), and (iii), one could proceed as follows. Rewrite his Eq. 7 as

$$p(a, \lambda, b, \mu, c, \nu) = \sum_{A,B} AB\,\{A \mid a, \lambda, c, \nu\}\,\{B \mid b, \mu, c, \nu\}.$$

Multiplying both sides by $\{\lambda \mid \nu, a, c\}\,\{\mu \mid \nu, b, c\}$ and integrating we obtain

$$P(a, b, c, \nu) = \bar{A}(a, c, \nu)\,\bar{B}(b, c, \nu),$$

where

$$\bar{A}(a, c, \nu) = \int \sum_{A} A\,\{A \mid a, \lambda, c, \nu\}\,\{\lambda \mid \nu, a, c\}\, d\lambda$$

$$\bar{B}(b, c, \nu) = \int \sum_{B} B\,\{B \mid b, \mu, c, \nu\}\,\{\mu \mid \nu, b, c\}\, d\mu.$$

Then Bell's argument (9) — (16) goes through, using (iii), without any friction at step (14).
3. The objection which we have raised against assertion (iii) holds equally well against an assuption made immediately after Eq. (4.2) in an earlier paper of Bell's [in *Proceedings of the International School of Physics "Enrico Fermi", Course IL, Varenna 1970* (Academic Press 1971), p. 178] and against the form of the joint probability distribution assumed for variables λ, λ', λ'' in a paper of Shimony [In *Logic, Methodology and Philosophy of Science,* P. Suppes, et al., eds., North-Holland Publishing Company, 1973, p. 567].
4. It is interesting to note that Bell's footnote 10 in his 1970 Varenna paper (see our previous footnote for reference) can be understood as an anticipation of a proof along the lines of that of Clauser and Horne.
5. See A. Shimony, "Scientific Inference", in *The Nature and Function of Scientific Theories,* ed. R. Colodny, (U. of Pittsburgh Press, Pittsburgh, 1971).

J.S. Bell - *Free Variables and Local Causality*

It has been argued[1] that quantum mechanics is not locally causal and cannot be embedded in a locally causal theory. That conclusion depends on treating certain experimental parameters, typically the orientations of polarization filters, as free variables. Roughly speaking it is supposed that an experimenter is quite free to choose among the various possibilities offered by his equipment. But it might be that this apparent freedom is illusory. Perhaps experimental parameters and experimental results are both consequences, or partially so, of some common hidden mechanism. Then the apparent non-locality could be simulated.

This possibility is the starting point of a paper by Clauser, Horne and Shimony[2] (CHS hereafter), which is valuable in particular for a careful mathematical formulation of the assumption which excludes such a conspiracy. In this connection they severely criticize my own "theory of local beables"[1] (B hereafter). Much of their criticism is perfectly just. In B there were jumps[3] in the argument, and the assumption in question was not stated at the appropriate place, but only later and inadequately. However, I do not agree with CHS that this assumption, when carefully formulated, is an unreasonable one.

I will organize these remarks around the three phrases in which I belatedly formulated the hypothesis in B, Section 8.

1) *"It has been assumed that the settings of instruments are in some sense free variables . . ."*

For me this means that the values of such variables have implications only in their future light cones. They are in no sense a record of, and do not give information about, what has gone before. In particular they have no implications for the hidden variables ν in the overlap of the backward light cones:

$$\{\nu \mid a, b, c\} = \{\nu \mid a', b, c\} = \{\nu \mid a, b', c\} = \{\nu \mid a', b', c\} \quad (1)$$

This, as explained by CHS, is what is used in the mathematical analysis. The bracket symbol denotes the probability of particular values ν given particular values a, b, c where c lists non-hidden variables in the overlap of the backward light cones of two instruments, and a and b list non-hidden variables in the remainders of those light cones. The lists a and a' are supposed to differ in the setting of the first instrument, while b and b' are supposed to differ in the setting of the second instrument.

Note that instead of (1) CHS write, probably interpreting the symbols a little differently

$$\{\nu \mid a, b, c\} = \{\nu \mid c\}$$

With my notation, where a and b are lengthy lists of variables describing the situation outside the overlap, this would be much stronger than (1) — and not reasonable at all.

2) "*... say at the whim of experimenters ...*»

Here I would entertain the hypothesis that experimenters have free will. But according to CHS it would not be permissible for me to justify the assumption of free variables "by relying on a metaphysics which has not been proved and which may well be false". Disgrace indeed, to be caught in a metaphysical position! But it seems to me that in this matter I am just pursuing my profession of theoretical physics.

I would insist here on the distinction between analyzing various physical theories, on the one hand, and philosophising about the unique real world on the other hand. In this matter of causality it is a great inconvenience that the real world is given to us once only. We cannot know what would have happened if something had been different. We cannot repeat an experiment changing just one variable; the hands of the clock will have moved, and the moons of Jupiter. Physical theories are more amenable in this respect. We can *calculate* the consequences of changing free elements in a theory, be they only initial conditions, and so can explore the causal structure of the theory. I insist that B is primarily an analysis of certain kinds of physical theory.

A respectable class of theories, including contemporary quantum theory as it is practised, have "free" "external" variables in addition to those internal to and conditioned by the theory. These variables are typically external fields or sources. They are invoked to represent experimental conditions. They also provide a point of leverage for "free willed experimenters", if reference to such hypothetical metaphysical entities is permitted. I am inclined to pay particular attention to theories of this kind, which seem to me most simply related to our everyday way of looking at the world.

Of course there is an infamous ambiguity here, about just what and where the free elements are. The fields of Stern-Gerlach magnets could be treated as external. Or such fields and magnets could be included in the quantum mechanical system, with external agents acting only on external knobs and switches. Or the external agents could be located in the brain of the experimenter. In the latter case the sitting of the instrument is *not* itself a free variable. It is only more or less closely correlated with one, depending on how accurately the experimenter effects his intention. As he puts out his hand to the knob, his hand may shake, and may shake in a way influenced by the variables ν. Remember, however, that the disagreement between locality and quantum mechanics is large — up to a factor of $\sqrt{2}$ in a certain sense. So some hand trembling can be tolerated without much change in the conclusion. Quantification of this would require careful epsilonics.

3) *"... or at least not determined in the overlap of the backward light cones"*

Here I must concede at once that the hypothesis becomes quite inadequate when weakened in this way. The theorem no longer follows. I was mistaken.

At this point I had in mind the possibility of exploiting the freedom, in conventional physical theories, of initial conditions. I am now embarrassed not only by the inadequacy of this particular phrase in the hypothesis, but also by the necessity of paying attention in such a study to the creation of the world[4].

Let me instead then weaken the hypothesis in a different and more practical way.

4) *"... or at least effectively free for the purpose at hand."*

Suppose that the instruments are set at the whim, not of experimental physicists, but of mechanical random number generators. Indeed it seems less impractical to envisage experiments of this kind[5], with space-like separation between the outputs of two such devices, than to hope to realize such a situation with human operators. Could the outputs of such mechanical devices reasonably be regarded as sufficiently free for the purpose at hand? I think so.

Consider the extreme case of a "random" generator which is in fact perfectly deterministic in nature and, for simplicity, perfectly isolated. In such a device the complete final state perfectly determines the complete initial state — nothing is forgotten. And yet for many purposes, such a device is precisely a "forgetting machine". A particular output is the result of combining so many factors, of such a lengthy and complicated dynamical chain, that it is quite extraordinarily sensitive to minute variations of any one of many initial conditions. It is the familiar paradox of classical statistical mechanics that such exquisite sensitivity to initial conditions is practically equivalent to complete forgetfulness of them. To illustrate the point, suppose that the choice between two possible outputs, corresponding to a and a', depended on the oddness or evenness of the digit in the millionth decimal place of some input variable. Then fixing a or a' indeed fixes something about the input — i.e., whether the millionth digit is odd or even. But this peculiar piece of information is unlikely to be the vital piece for any distinctively different purpose, i.e., it is otherwise rather useless. With a physical shuffling machine, we are unable to perform the analysis to the point of saying just what peculiar feature of the input is remembered in the output. But we can quite reasonably assume that it is not relevant for other purposes. In this sense the output of such a device is indeed a sufficiently free variable for the purpose at hand. For this purpose the assumption (1) is then true enough, and the theorem follows.

Arguments of this kind are advanced by CHS in defending the corresponding assumption in the Clauser-Horne analysis. I do not know why they should be considered less relevant here.

Of course it might be that these reasonable ideas about physical randomizers are just wrong — for the purpose at hand. A theory may appear in which such conspiracies inevitably occur, and these conspiracies may then seem more digestible than the non-localities of other theories. When that theory is announced I will not refuse to listen, either on methodological or other grounds. But I will not myself try to make such a theory.

J.S. Bell, CERN — Genève

REFERENCES

(1) J.S. Bell, Epistemological Letters *9,* March 1976, and references therein (17.0).

(2) A. Shimony, M.A. Horne and J.F. Clauser, Epistemological Letters *13,* octobre 1976 (17.1).

(3) In particular CHS complain of a difficulty in connection with (14) and (15) in B. What is missing there is the remark that on the right-hand sides the averaging is over μ and μ', and λ and λ', separately. The operation is, explicitly,

$$\int d\lambda\, d\lambda'\, d\mu\, d\mu'\, d\nu \{\lambda \mid a, c, \nu\}\{\lambda' \mid a', c, \nu\}\{\mu \mid b, c, \nu\}\{\mu' \mid b', c, \nu\}\{\nu \mid a, b, c\}$$

According to Eq. (1) above a and/or b in the last factor may be replaced by a' and/or b' respectively. As applied for example to

$$p(\lambda, a), (\mu, b), (\nu, c)$$

which does not depend on λ' or μ', two integrations are trivial, leaving

$$\int d\lambda\, d\mu\, d\nu \{\lambda \mid a, c, \nu\}\{\mu \mid b, c, \nu\{\nu \mid a, b, c\}$$

$$\equiv \int d\lambda\, d\mu\, d\nu \{\lambda \mid a, b, \mu, c, \nu\}\{\mu \mid a, b, c, \nu\}\{\nu \mid a, b, c\}$$

(using locality)

$$\equiv \int d\lambda\, d\mu\, d\nu \{\lambda, \mu, \nu \mid a, b, c\}$$

which is the averaging involved in defining P (a, b, c).
However, I agree with CHS that an earlier style[6] averaging over λ and μ before forming the inequality, is simpler.

(4) The invocation in Ref. 1) of a *complete* account of the overlap of backward light cones is embarrassing in a related way, whether going back indefinitely or to a finite creation time — which might, by the way, have even been a creation *point,* with all backward light cones confused. R.P. Feynman in particular objected to the concept of a complete history being involved. In a more careful discussion the notion of completeness should perhaps be replaced by that of sufficient completeness for a certain accuracy, with suitable epsilonics.

(5) Important progress in this direction is being made by A. Aspect, Physical Review *D14,* 1944 (1976).

(6) J.S. Bell, in Proceedings of the International School of Physics "Enrico Fermi", Course IL, Varenna 1970.

Abner Shimony - *Reply* to Bell 17. 3*

Bell's answer[1] to criticisms[2] by Shimony, Clauser, and Horne (CHS) of his earlier paper[3] has greatly diminished the distance between their respective positions. At risk of exhibiting what Freud called "narcissism of small differences" I shall comment on a crucial passage in Bell's answer. He says: "Consider the extreme case of a "random" generator which is in fact perfectly deterministic in nature - and, for simplicity, perfectly isolated. In such a device the complete final state perfectly determines the complete initial state - nothing is forgotten. And yet for many purposes, such a devise is precisely a "forgetting machine". A particular output is the result of combining so many factors, of such a lengthy and complicated dynamical chain, that it is quite extraordinarily sensitive to minute variations of any one of many initial conditions. It is the familiar paradox of classical statistical mechanics that such exquisite sensitivity to initial conditions is practically equivalent to complete forgetfulness of them. To illustrate the point, suppose that the choice between two possible outputs, corresponding to a and a', depended on the oddness or evenness of the digit in the millionth decimal place of some input variable. Then fixing a or a' indeed fixes something about the input - i. e., whether the millionth digit is odd or even. But this peculiar piece of information is unlikely to be the vital piece for any distinctively different purpose, i. e., it is otherwise rather useless. With a physical shuffling machine, we are unable to perform the analysis to the point of saying just what peculiar feature of the input is remembered in the output. But we can quite reasonably assume that it is not relevant for other purposes. In this sense the output of such device is indeed a sufficiently free variable for the purpose at hand. For this purpose the assumption (1) is then true enough, and the theorem follows. "

This passage is excellent up to the last sentence. The last sentence, however, seems to me to be a non-sequitur, unless the phrase "true enough" is interpreted with extreme latitude. Suppose that in Bell's idealized example of a deterministic and completely isolated generator the set of hidden variables ν is such that a would be generated and a' not. Then for all b and c

$$\{ \nu \mid a',b,c \} = 0$$

where the expressions on the left hand side denotes the probability (or probability density) that the hidden variables have the value ν , on condition that

* Work supported by the National Science Foundation

the non-hidden variables are a', b and c. On the other hand, for at least some b and c

$$\{\nu \mid a,b,c\} \neq 0.$$

Hence Bell's assumption (1), which asserts the equality of $\{\nu \mid a,b,c,\}$ and $\{\nu \mid a',b,c\}$ is false.

One can guess what Bell means by "true enough" from several sentences before the last in the passage quoted. Suppose K is some much less comprehensive feature of the generator than the complete set of hidden variables. Then from knowledge of K no inference can be drawn about whether a or a' some other output will be generated. If so, then it is reasonable to assume that

$$\{K \mid a,b,c\} = \{K \mid a',b,c\}.$$

(Rigorous examples in which this assumption is true are provided by dynamical systems having the "mixing property".[4] But this assumption is essentially the same as assumption (2) on p. 5 of Ref. 2. Thus the distance between the positions of Bell and of CHS seems to have converged to zero, but the latter can still claim to have articulated the common position with greater clarity.

There is an objection which Bell could have brought against CHS. They express a preference for the Clauser-Horne derivation[5] of Bell's inequalities over all other derivations, on grounds of generality and plausibility of assumptions. In my opinion, however, they should have recognized that Bell's derivation of 1971,[6] though very different in argument, proceeds from essentially the same assumptions. Bell assumes that the expectation value $\bar{A}$ depends only on the apparatus setting a and the hidden variables λ, and similarly for $\bar{B}$. Clauser and Horne assume that the probability p_1 of detection of emission 1 depends only upon a and λ, and similarly for the probability of detection of emission 2. When one recognizes the correct operational connections between the expectation values of observables and detection probabilities, the equivalence of these assumptions becomes clear.

A final remark concerns an entirely different point: namely, Bell's term "beable". The suffix "able" etymologically refers to a potentiality. An observable of a system is a property which can be observed, even though it may actually be the case that no one has observed it. But Bell's criterion for applying the term "beable" to things is that "they are there" (Ref. 3, p. 11); no potentiality is involved. The term "existent" would have been more accurate than his neologism. I hope that Gresham's law will not be confirmed in the present case.

A. Shimony, Boston University

REFERENCES

(1) J.S. Bell, Epistemological Letters, 15, Febr. 1977
(2) A. Shimony, M.A. Horne, and J.F. Clauser, Epistemological Letters 13, October 1976
(3) J.S. Bell, Epistemological Letters 9, March 1976
(4) See, for example, J. Lebowitz and O. Penrose, Physics Today *26*, No. 2, 23 (1973).
(5) J.F. Clauser and M.A. Horne, Phys. Rev. D10, 526 (1974)
(6) J.S. Bell, in *Foundations of Quantum Mechanics:* Proceedings of the Int. School of Physics "Enrico Fermi", Course IL, Varenna 1970, ed. by B. d'Espagnat.

SELECTIVE BIBLIOGRAPHY OF RECENT WORKS ON HIDDEN VARIABLES

I. *Surveys and Collections*

Pipkin, F.M., Advances in Atomic and Molecular Physics *14,* 281-340 (1978).

Clauser, J.F. and Shimony, A., Reports on Progress in Physics *41,* 1881-1927 (1978).

Selleri, F. and Tarozzi, G., Rivista del Nuovo Cimento *4,* no. 2, 1-53 (1981).

d'Espagnat, B., Physics Reports *110,* 202-264 (1984).

Wheeler, J.A. and Zurek, W.H., eds. *Quantum Theory and Measurement* (Princeton University Press, Princeton, N.J., 1983). Section III of this comprehensive anthology contains nine papers concerning hidden variables theories and Bell's theorem. There is an annotated bibliography.

Bell, J.S., Foundation of Physics *12,* 989-999 (1983). This is a valuable historical document, presenting the background and the heuristics of Bell's theorem in a way that is not available elsewhere.

Flato, M., Maric, Z., Milojevic, A., Sternheimer, D., and Vigier, J.-P., eds., *Quantum Mechanics, Determinism, Causality and Particles* (Reidel, Dordrecht, 1976).

II. *New Derivations of Inequalities of Bell's Type*

Garuccio, A. and Selleri, F., Foundations of Physics *10,* 209-216 (1980).

Mermin, N.D., Physical Review D *22,* 356-361 (1980).

Ogren, M., Physical Review D *27,* 1766-1773 (1983).

Garg, A. and Mermin, N.D., Foundations of Physics *14,* 1-39 (1984).

Mermin, N.D., Physics Today *38,* no. 4, 38-47 (1985).

III. *Analysis of Locality*

Fine, A., in *PSA 1980,* Asquith, A. and Giere, R., eds. (Philosophy of Science Association, E. Lansing, Mich., 1980), 535, with critical comments by A. Shimony.

Jarrett, J., Nous *18,* 569-589 (1984).

Shimony, A., in *Foundations of Quantum Mechanics in the Light of New Technology,* Kamefuchi, S. *et al,* eds. (Physical Society of Japan, Tokyo, 1984), 225-230.

Bohm, D. and Hiley, B., *ibid.,* 231-232.

IV. *Recent Experiments*

Aspect, A., Grangier, P., and Roger, G., Physical Review Letters *49,* 91-94 (1982).

Aspect, A., Dalibard, J., and Roger, G., Physical Review Letters *49,* 1804-1807 (1982). This is the most important of all the recent results concerning Bell's theorem. The authors tested a

Bell-type inequality, using an apparatus which effectively selected the orientations of polarization analyzers for each of a pair of photons while the photons were in flight.

V. *Significance of Bell's theorem and Related Experiments*

Fine, A., Physical Review Letters *48,* 291-295 (1982) and Journal of Mathematical Physics *23,* 1306-1310 (1982). These papers argue that the essential content of Bell's theorem concerns the non-existence of certain joint probability distributions, and that locality is a peripheral issue.

Shimony, A., British Journal for the Philosophy of Science *35,* 25-45 (1984). This paper contains a critique of the foregoing.

Marshall, T.W., Santos, E., and Selleri, F., Physics Letters *98A,* 5 (1983).

Piccioni, O., Bowles, P., Enscoe, C., Garland, R., and Mehlhop, W., in *Open Questions in Quantum Physics,* Tarozzi, G. and van der Merwe, A., eds. (Reidel, Dordrecht, 1984).

Reprinted with permission from *Physics Letters*, 90A, J. Summhammer, G. Badurek, H. Rauch, and U. Kischko, "Explicit Experimental Verification of Quantum Spin-State Superposition," pp. 110-112, ©1982

Volume 90A, number 3 PHYSICS LETTERS 28 June 1982

EXPLICIT EXPERIMENTAL VERIFICATION OF QUANTUM SPIN-STATE SUPERPOSITION ☆

J. SUMMHAMMER, G. BADUREK, H. RAUCH
Atominstitut der Österreichischen Universitäten, A-1020 Vienna, Austria

and

U. KISCHKO
Institut für Physik, Universität, D-46 Dortmund, Fed. Rep. Germany
and Institut Laue–Langevin, F-38042 Grenoble, France

Received 2 April 1982

An interferometric experiment with polarized neutrons is described that demonstrates the basic principles of coherent spin-state superposition in an explicit way.

The principle of linear superposition of probability amplitudes of coherent quantum states is one of the basic foundations of quantum mechanics. However, as far as we know, no explicit experimental verification or, more precisely, demonstration of this principle for spin states has been presented until now. In 1963 Wigner [1] proposed a Gedanken experiment where he imagined a double Stern–Gerlach arrangement to separate and subsequently recombine coherently the z-spin components of an x-spin particle. Zeilinger [2] suggested the realization of such an experiment by means of neutron interferometry. The detailed theoretical framework by means of which the measured results can be interpreted conveniently is treated by Eder and Zeilinger [3].

In this letter we report on an experiment with polarized neutrons that was performed at the perfect crystal neutron interferometer setup (D18) at the high flux reactor Grenoble [4]. The experimental arrangement is sketched schematically in fig. 1. Nearly 100% neutron polarization is achieved by combination of the spin-dependent double refraction of the

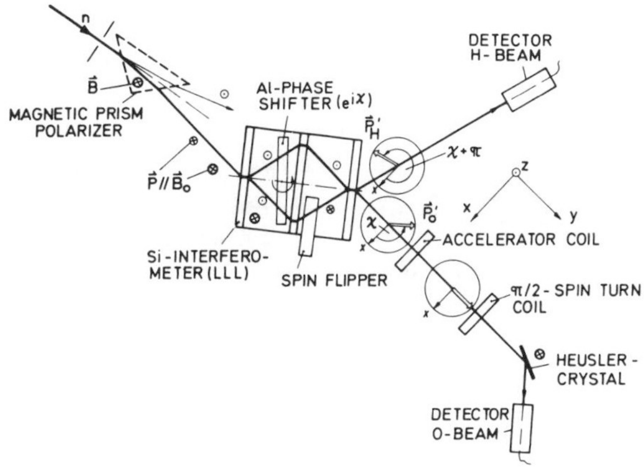

Fig. 1. Schematic of the experimental arrangement.

incident unpolarized beam at the boundaries of a prismatically shaped magnetic field region and of the intrinsically extremely narrow angular reflection width of the double-crystal arrangement [5]. Proper angular positioning of the interferometer crystal thus allows both of the two coherent wave trains within this triple-Laue-case (LLL) interferometer to be either in the spin "up" or spin "down" state with respect to the z-axis. This axis is chosen to coincide with the direction of a magnetic guide field that is produced by a Helmholtz coil pair covering the whole setup. Since

☆ Financially supported by the Austrian Fonds zur Förderung der wissenschaftlichen Forschung (Projekt no. 4230) and by The Bundesminister für Forschung und Technologie, Bonn (F. no. 03-41 E 09 P).

 0 031-9163/82/0000–0000/$02.75

Volume 90A, number 3 PHYSICS LETTERS 28 June 1982

the two partial waves are well separated in space it is possible to invert the spin state of one of them by means of a miniaturized dc-coil spin flip device, leaving the other one unaffected. To avoid an unnecessarily large loss of coherence and hence interference contrast because of thickness inhomogeneities the windings of this coil were made partially of a Nb–V alloy whose index of refraction for neutrons is practically 1. Careful thermal isolation was necessary as well to prevent untolerable temperature gradients within the interferometer crystal. More details will be given in a forthcoming paper [6]. Rotation of a plan-parallel sample within the two coherent beams introduces a nuclear phase shift proportional to the path difference ΔD within the material. The corresponding phase difference is then given as $\chi = -N\lambda b_c \Delta D$, where N is the number of nuclei per unit volume, λ is the neutron wave length and b_c is the coherent scattering length of the phase shifter material. Behind the third crystal plate of the interferometer a forward (0) and deviated (H) diffracted beam emerge which are coherent superpositions of the partial waves propagating inside the interferometer. Because of particle number and angular momentum conservation both their intensity and their polarization are mutually complementary to each other. It was shown theoretically [3] and verified experimentally [7] that the two emerging beams are polarized even in the case of unpolarized incident neutrons if nuclear phase-shift and magnetic spin-rotation are applied simultaneously within the interferometer. According to the notations given in ref. [3] the combined action of phase-shift and spin-rotation is described by the unitary operator

$$U = \exp(\mathrm{i}\chi)\exp(-\mathrm{i}\boldsymbol{\sigma}\cdot\boldsymbol{\alpha}/2)$$
$$= \exp(\mathrm{i}\chi)(\cos\alpha/2 - \mathrm{i}\boldsymbol{\sigma}\cdot\hat{\boldsymbol{\alpha}}\sin\alpha/2)\,, \qquad (1)$$

where $\boldsymbol{\sigma}$ is the Pauli spinor operator and $\boldsymbol{\alpha} = \alpha\hat{\boldsymbol{\alpha}}$ is the vector of rotation. The intensity and polarization of the forward beam, for example, are then obtained formally by

$$I = \Psi_0^{\mathrm{I}+}(1 + U^+)(1 + U)\Psi_0^{\mathrm{I}} \qquad (2)$$

and

$$P = \Psi_0^{\mathrm{I}+}(1 + U^+)\boldsymbol{\sigma}(1 + U)\Psi_0^{\mathrm{I}}/I\,, \qquad (3)$$

where Ψ_0^{I} denotes the wave function of one of the two (identical) partical 0-beam components inside of an ideal empty interferometer. Of particular importance for our experiment is the case of spin state inversion, which in our specific situation is a rotation of the spin pointing initially in the $+z$ direction by an angle $\alpha = \pi$ around the x-axis. Here the unitary operator reduces to

$$U = -\mathrm{i}\sigma_x \exp(\mathrm{i}\chi)\,, \qquad (4)$$

and it can be shown that in this case the polarization of the forward beam results as

$$\boldsymbol{P} = \sin\chi\,\boldsymbol{e}_x - \cos\chi\,\boldsymbol{e}_y\,, \qquad (5)$$

whereas its intensity is a constant independent of the value of the nuclear phase shift χ.

According to this last equation the final polarization lies within the x–y plane though both beams inside the interferometer were polarized into $+z$ and $-z$ direction, respectively. This means that the outgoing wave has properties which neither of the two interfering waves had and therefore could never be explained as a classical mixture of the superposed spin states.

To verify eq. (5) we have demonstrated that the polarization of the forward beam has indeed zero z-component and rotates entirely within the x–y plane as a

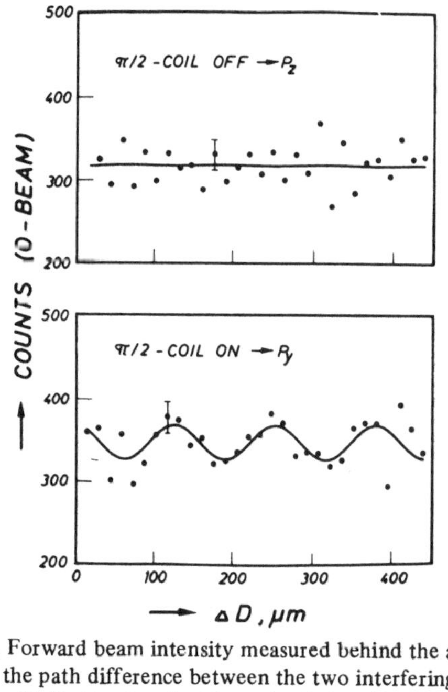

Fig. 2. Forward beam intensity measured behind the analyzer versus the path difference between the two interfering partial waves with (bottom) and without (top) the $\pi/2$-spin turn coil in action. The full lines correspond to least-squares fits of sinusoidal curves to the experimental values.

Volume 90A, number 3 PHYSICS LETTERS 28 June 1982

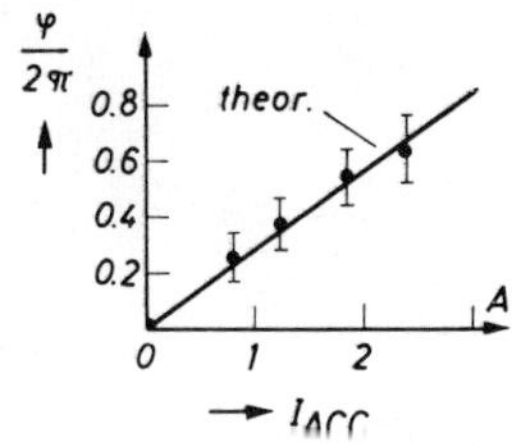

Fig. 3. Observed and calculated phase shift of the intensity oscillation pattern as a function of the dc current fed into the accelerator coil.

function of χ. This was achieved by placing a so-called $\pi/2$-spin turn coil, which mutually exchanges the y- and z-components of the polarization vector, in front of the magnetic Heusler single crystal analyzer which in turn was aligned in the z-direction. In fig. 2 the detected intensity behind the analyzer is plotted as a function of the path difference of the two interfering beams with and without the action of the spin turn coil. In spite of the relatively large statistical error of the measured values, which is essentially due to the dominance of the noninterfering part of the intensity, it is readily seen that no phase-shift dependent coherent intensity variation occurs if the z-component of the polarization is analyzed. The y-component, on the other hand, exhibits a significant modulation to which a sinusoidal curve fits quite well.

In order to compensate for the spatial precession of the polarization around the magnetic guide field $\boldsymbol{B}_0$, a so-called accelerator coil was inserted into the 0-beam path which allowed for a controlled additional phase shift $\varphi = \gamma B_{\mathrm{acc}} d/v$ of the interference pattern by means of local variation of magnetic field strength. Here $\boldsymbol{B}_{\mathrm{acc}}$ (parallel to $\boldsymbol{B}_0$) and d are the field and the thickness, respectively, of the accelerator coil, $\gamma = 1.83 \times 10^8$ rad s^{-1}T^{-1} is the gyromagnetic ratio of the neutrons and v their velocity. For each angular position of the nuclear phase shifter the detector response was recorded for a series of dc currents fed into the coil. As it should be, the shift of the intensity modulation pattern was found to depend strictly linearly on the current values. This result is shown in fig. 3.

In a forthcoming paper [6] we will present a more detailed description of the experiment as well as a more elaborate analysis of the measurements. These include in fact also the case of partial absorption of one wave train inside the interferometer and the application of a radio frequency spin flipper instead of the static one described here. Furthermore we will discuss in terms of dynamical diffraction theory the influence of magnetic fields and spin state inversion on the reflectivity of the crystal plates of the interferometer.

This work was performed as part of the Vienna-Dortmund group cooperation of the Institute Laue–Langevin, Grenoble. The authors wish to thank Prof. G. Eder and Prof. A. Zeilinger for valuable discussions.

References

[1] E.P. Wigner, Am. J. Phys. 31 (1963) 6.
[2] A. Zeilinger, in: Neutron interferometry, eds. U. Bonse and H. Rauch (Oxford U.P., London, 1979) p. 265.
[3] G. Eder and A. Zeilinger, Nuovo Cimento 34B (1976) 76.
[4] W. Bauspiess, U. Bonse and H. Rauch, Nucl. Instrum. Methods 157 (1978) 495.
[5] G. Badurek, H. Rauch, A. Wilfing, U. Bonse and W. Graeff, J. Appl. Cryst. 12 (1979) 186.
[6] J. Summhammer, G. Badurek, H. Rauch, U. Kischko and A. Zeilinger, Phys. Rev. B, to be published.
[7] G. Badurek, H. Rauch, A. Zeilinger, W. Bauspiess and U. Bonse, Phys. Rev. D14 (1976) 1177.

PHYSICAL REVIEW D VOLUME 9, NUMBER 4 15 FEBRUARY 1974

Experimental distinction between the quantum and classical field-theoretic predictions for the photoelectric effect*

John F. Clauser

Department of Physics and Lawrence Berkeley Laboratory, University of California, Berkeley, California 94720

(Received 30 October 1973)

We have measured various coincidence rates between four photomultiplier tubes viewing cascade photons on opposite sides of dielectric beam splitters. This experimental configuration, we show, is sensitive to differences between the classical and quantum field-theoretic predictions for the photoelectric effect. The results, to a high degree of statistical accuracy, contradict the predictions by any classical or semiclassical theory in which the probability of photoemission is proportional to the classical intensity.

INTRODUCTION

It is commonly believed that experimental observations of the photoelectric effect establish the existence of uniquely quantum-mechanical properties of the electromagnetic field. Various classic experiments, coupled with the notion of microscopic energy conservation, are usually cited to establish this claim.[1] Unfortunately the insistence upon microscopic energy conservation amounts to an auxiliary criterion, which for a classical field theory (CFT) is inherently ambiguous. The quantum-mechanical energy of a photon, $h\nu$, is experimentally relevant to the photoelectric effect, determining the kinetic energy of the ejected electrons. This insistence, on the other hand, demands that the classical field energy $\int(E^2+H^2)dV/8\pi$ be equal to $h\nu$ and be simultaneously conserved. The classical Maxwell equations contain no constraint that these energies be equal, as a quantum field theory (QFT) does.[2] This demand is, in fact, unreasonable for a classical field theory. It is therefore also unreasonable to use this constraint as a basis for an experimental distinction between the theories. With equal justification one might say that these experiments disprove microscopic energy conservation during the photoelectric process while upholding CFT. The above belief was finally shown to be totally unfounded when it was demonstrated that the above observations can be quantitatively accounted for by a semiclassical radiation theory in which the electromagnetic field is left unquantized.[3] The basic elements of this theory have since been used as a skeleton for the more recent and widely discussed neoclassical radiation theory (NCT) of Jaynes, Crisp, and Stroud.[4] In both of these theories it is hypothesized that the classical Maxwell equations describe the free electromagnetic field, and that this field never needs to be quantized to account for experimental observations. Previous experimental observations of the photoelectric effect, in and of themselves, are in agreement with this hypothesis, and do not appear to necessitate quantum-mechanical properties for the radiation field.

In 1955, following Schrödinger's suggestion, Ádám, Jánossy, and Varga[5] (AJV) searched for anomalous coincidences in a partially collimated beam of light.[5a] Jauch,[6] in his discussions of the foundations of quantum mechanics, has recently emphasized the importance of this experiment and an associated one performed by Jánossy and Náray[7] in establishing the existence of a wave-particle duality for photons. Moreover, the arguments of AJV and Jauch do not rely on energy conservation (although other assumptions are needed for their specific scheme) and as such are not subject to the above criticism. Attention is naturally called to this experiment by the recent discussions of semiclassical theories, in hopes that it might provide an additional aspect of the photoelectric effect upon which the predictions of CFT and QFT differ.

In this paper we will show that the actual values of the parameters for the arrangement of AJV (and subsequent similar experiments) unfortunately were insufficient to make that experiment conclusive. We then report new experimental results which are conclusive. Our measurements involved a comparison of various twofold coincidence rates between four photomultiplier tubes viewing cascade optical photons emitted by the same source through various beam splitters. We further show that this configuration is sensitive to differences between the QFT and CFT predictions for this effect without additional assumptions, such as those required by AJV. The results, to high statistical accuracy, contradict the predictions of any classical or semiclassical radiation theory in which the probability of photoemission is proportional to the classical field intensity. This includes, for example, NCT. Our experiment thus resurrects the photoelectric effect as a phenomenon requiring quantization of the electromagnetic field.

It is noteworthy that Aharonov *et al.*[8] presented a scheme similar to that of AJV as a *Gedanken-experiment,* while noting a paucity of actual experimental distinctions between CFT and QFT. The CFT prediction for our experiment follows reasoning similar to that by Titulaer and Glauber,[9] who discussed constraints applicable to CFT which demarcate a boundary between CFT and the more general QFT descriptions of the electromagnetic field.

In what follows we first contrast the CFT and QFT predictions for a single photon falling on a half-silvered mirror. We next discuss previous relevant experiments, contrast these with our own experimental scheme, and show that of these only ours provides the desired distinction. Finally we describe the apparatus and present the results.

PREDICTIONS FOR A SINGLE PHOTON FALLING ON A HALF-SILVERED MIRROR

In this section we review the arguments by AJV and Jauch. Consider the light emitted by a single atomic decay falling on a half-silvered mirror. During the decay a wave train (packet) of electromagnetic radiation is emitted. Suppose that it impinges upon a beam-splitting mirror, and that the two resultant wave trains are directed to two independent photomultipliers labeled γ_A and γ_B. We desire the QFT prediction for the γ_A-γ_B coincidence rate. A simpler problem to consider first involves only the source atom and a second atom in one photocathode. We need the probability amplitude that, following deexcitation of the source atom, the second atom will become excited (or ionized). This has been obtained by Fermi[10] and Fano,[11] using the Wigner-Weisskopf approximation. The inclusion of a third atom in a second photocathode is then straightforward. Denote by S, A, and B, respectively, the ground states of the source atom and the two detector atoms, and by S^*, A^*, and B^* the corresponding excited or ionized states of these atoms. Initially the source atom is excited, and the two detector atoms are in their ground states; hence

$$|i\rangle = |S^*, A, B, 0_1, \ldots, 0_j, \ldots\rangle .$$

The remaining indices of the ket designate the state of the radiation field modes. The final state then has the form

$$\begin{aligned}|f\rangle = {} & U_A|S, A^*, B, 0_1, \ldots, 0_j, \ldots\rangle \\ & + U_B|S, A, B^*, 0_1, \ldots, 0_j, \ldots\rangle \\ & + U_S|S^*, A, B, 0_1, \ldots, 0_j, \ldots\rangle \\ & + \sum_j U_j|S, A, B, 0_1, \ldots, 1_j, \ldots\rangle . \end{aligned} \qquad (1)$$

The various U_i can be evaluated from formulas found in Refs. 10 and 11. Thus QFT predicts that an observation will find at most one of the detector atoms ionized; i.e., coincident responses will occur only at the random accidental rate, induced by emissions from two different excited source atoms.[12]

Next we consider the same system from the CFT viewpoint. Our basic assumptions for this are twofold: (1) The electromagnetic field is described by the classical (unquantized) Maxwell equations, and (2) the probability of photoionization at a detector is proportional to the classical intensity of the incident radiation. These two assumptions alone are sufficient for our purposes, and they are in evident agreement with experiment.[13] Since ionizations at the γ_A and γ_B phototubes are independent, but are induced by nearly identical classical pulses of light, for a given split wave train both tubes will have roughly the same probability for registering a count. This independence implies that the probability that both will respond to the split wave train is simply the product of the probabilities that each will respond. The nonzero value of this product implies the existence of an anomalous coincidence rate above the accidental background. The CFT prediction is thus in marked contrast with QFT prediction, the latter requiring no coincidences above the background level.[12]

The above argument may be summarized very simply. Consider a radiation field quantum-mechanically with only one photon present. If we bring this into interaction with two separated atoms we will never get more than one photoelectron. If on the other hand we represent this field classically, we find that there is a nonvanishing probability for finding two photoelectrons. The classical Maxwell field has within it the possibility of providing with some probability any number of photons. Hence experiments of the above variety can distinguish between the two theories.

Such then is the argument of AJV and Jauch. Here we have also the basis for the usual particle interpretation of photons. A particle must be either transmitted or reflected. Both may be done simultaneously only by a wave. We then see how these macroscopic features of "particle-like" objects arise from the QFT formalism.

PREVIOUS EXPERIMENTAL RESULTS

That a photon is not split in two by a beam splitter is certainly "old hat," and it may seem surprising that we have gone to the effort to test this prediction experimentally. What is in fact much more surprising is that evidently no such experimental test has heretofore been performed, and such tests are clearly of great importance. Here we briefly review previous relevant experimental results and show that none provides the desired distinction.

Since the original work of AJV many two-photon coincidence experiments have been done, some involving light beams split by a half-silvered mirror. These all fall into two basic categories—atomic-cascade observations and Brown-Twiss-effect observations. Excellent reviews of these topics have been presented by Camhy-Val and Dumont[14] and Mandel and Wolf,[15] respectively. Cascade-photon observations in their usual configuration are not suitable for the above test, since in these, two different unsplit photons are observed.

The AJV experiment, although intended as a test of the above scheme, actually served as a forerunner to the Brown-Twiss-effect experiments. Figure 1 reproduces a diagram of the experiment of AJV. In it they selected the light of a single spectral line with a monochromator, and focused it through a beam splitter onto two photomultipliers whose outputs drove a coincidence circuit.

Let us evaluate the magnitude of the expected anomalous coincidence rate. The CFT predictions for one and two photodetectors sharing the same field were discussed earlier by Mandel[16] from the above fundamental assumptions. Denote by $I(t)$ the instantaneous classical intensity incident simultaneously upon the γ_A and γ_B detectors due to their illumination by the whole source volume. The singles rates for the A and B detectors, averaged over their response time T, is given by

$$S_A = \alpha_A T^{-1}\int_{-T/2}^{T/2} \langle I(t+t')\rangle dt' ,$$
$$S_B = \gamma_B T^{-1}\int_{-T/2}^{T/2} \langle I(t+t')\rangle dt' , \qquad (2)$$

where α_A and α_B are measures of the detector efficiencies, and the angular brackets denote an ensemble average over the emitted intensities. Similarly, the average coincidence rate as a function of event separation τ is given by

$$C_{AB}(\tau) = \alpha_A \alpha_B T^{-1}\int_{-T/2}^{T/2}\int_{-T/2}^{T/2} \langle I(t+t')I(t+t''+\tau)\rangle dt'dt''. \qquad (3)$$

To obtain a model-independent prediction for the coincidence rate only from data on the singles rates does not appear possible, since (2) and (3) involve different averages of $I(t)$. AJV thus had to make various assumptions (assumptions which were unnecessary in the case of our own experiment). They tacitly assumed that

$$\left(\int_{-T/2}^{T/2} \langle I(t+t')\rangle dt'\right)^2 \approx \int_{-T/2}^{T/2}\int_{-T/2}^{T/2} \langle I(t+t')I(t+t''+\tau)\rangle\, dt'dt'' \qquad (4)$$

holds for each decay, when τ is the order of the decaying state lifetime. If then E pulses per sec-

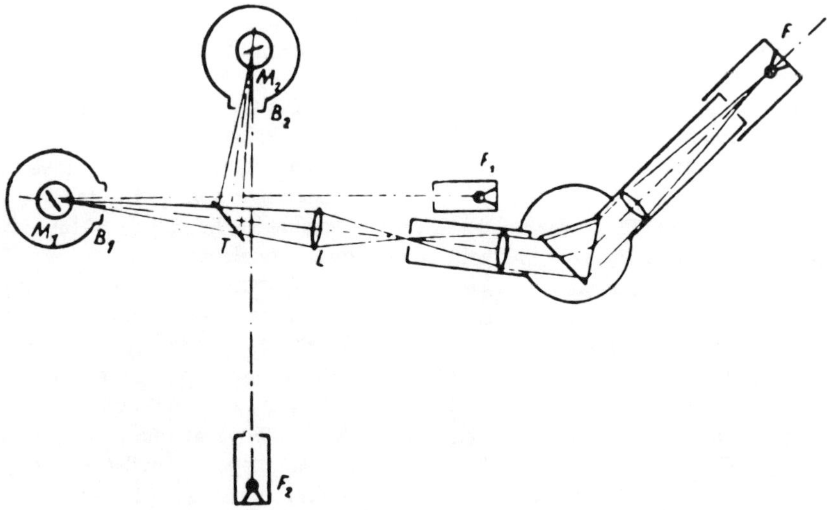

FIG. 1. Experimental arrangement of Ádám, Jánossy, and Varga. Light from source F is focused through a monochromator onto photomultipliers M_1 and M_2 via beam splitter T. (Figure after Ádám, Jánossy, and Varga.)

ond are emitted per unit time by a source, and if η is the average probability that a photomultiplier will yield a count, given an atomic decay, the count rate at that detector is

$$S = E\eta . \quad (5)$$

The expected anomalous coincidence rate predicted by the AVJ assumptions is then given approximately by

$$C \approx \eta^2 E . \quad (6)$$

Assuming negligible detector dark rates, the accidental coincidence background rate from which C must be distinguished is

$$A \approx \eta^2 E^2 2\tau_c , \quad (7)$$

where τ_c is the resolving time of the coincidence system. One can now calculate the integration time required to measure to a precision of N standard deviations the difference between the excess coincidence rate given by Eq. (6) and the zero excess rate predicted by QFT. Doing this we obtain

$$T_{\text{int}} \approx (1 + 4E\tau_c) N^2 \eta^{-2} E^{-1} , \quad (8)$$

which in the limit of high source rates takes the form

$$T_{\text{int}} \approx 4N^2 \tau_c / \eta^2 . \quad (9)$$

Thus the validity of their experiment rests directly upon the assumed or measured value of η: If it is too small, T_{int} will be too long and the experiment will see only the random accidental background. AJV measured their detector efficiencies by assuming that these were given by the formula

$$\eta = Rh\nu / W , \quad (10)$$

where R is the count rate obtained for a given beam of photons, and W is the power in the same beam measured bolometrically. They thus found $\eta = 1/300$. With a resolving time $\tau_c = 2.3$ μsec one calculates $T_{\text{int}} = 20.7$ sec for $N = 5$. From this reasoning AJV felt confident that they should have observed the anomalous coincidence rate, if it was present.

Let us reexamine from the CFT viewpoint the assumption tacitly contained in Eq. (10). Although the introductory arguments did not contain a requirement for energy conservation, AJV have unnecessarily reintroduced it with this assumption; this is in direct conflict with our fundamental assumptions for a CFT. In our derivation above, η is the probability for a detector response, given a source atom decay. Clearly a wavelike pulse emitted by a source atom will expand, in the worst case spherically, or at best with a radiation pattern having a preferred direction.[17] Much of this pulse will not enter the narrow acceptance solid angle subtended by the monochromator. Propagation will cause it to suffer an enormous decrease in intensity, commensurate with its expansion. Assuming macroscopic energy conservation on the average, the power W should then represent the total average power radiated by the source at the appropriate wavelength, not that which happens to be measured within the beam itself. The number calculated from Eq. (10) must be appropriately decreased by the fraction of the solid angle effectively subtended by the detectors. Other optical losses will decrease this number even further.

If we conservatively estimate from their diagram the solid-angle loss to be 1/400, their actual detector efficiency for spherically emitted wavelike pulses was undoubtedly less than 8×10^{-6}, in which case the required integration time for even $N = 1$ becomes $T_{\text{int}} \approx 1.3 \times 10^5$ sec. This is an order of magnitude longer than the duration of their experiment. Thus the experiment of Ádám, Jánossy, and Varga appears to be inconclusive when reexamined in this light.

A similar analysis applies to the experiments of Givens and of Brannen and Ferguson.[5a] In the x-ray coincidence experiment of Givens, the source solid angle viewed by the detector pair was $\approx 3.5 \times 10^{-5}$ sr, smaller than that of AJV. Combining this with his $\approx 15\%$ quantum efficiencies, we estimate the over-all detector efficiencies to be $\approx 2.1 \times 10^{-7}$ (neglecting the appreciable loss due to the beam-splitting crystal). Givens employed a resolving time of $\approx 1.7 \times 10^{-4}$ sec. From Eq. (9), we find then that an integration time of nearly 500 yr is required for this apparatus to produce results with a confidence level corresponding to just one standard deviation. Similar reasoning finds the actual integration time of Brannen and Ferguson deficient by a factor $\approx 1.7 \times 10^5$. These experiments are thus likewise inconclusive for deciding the above question.

Finally let us consider experiments of the Brown-Twiss variety. These experiments have a configuration basically the same as that of AJV. Because of the nature of this effect, however, all existing data have been accumulated with detectors subtending extremely small solid angles, much smaller even than those of AJV. From Eq. (9) we see that the required integration time scales with the inverse square of the detector solid angles; hence it would be hopeless to try to search for the above anomalous coincidence rate with such arrangements. Furthermore, in these experiments, the Brown-Twiss effect itself would tend to mask the effect we seek. In summary, then, none of the above experiments can provide the desired distinction.

EXPERIMENTAL SCHEME REQUIRING NO ADDITIONAL ASSUMPTIONS

The above discussion indicates that an observation of the anomalous coincidences predicted by a CFT requires highly efficient photodetectors. However, even if AJV had had the required efficiency and integration time, their experimental arrangement necessitated assumptions concerning the various field averages, and hence assumed a basic model for the emission mechanism. Since no universally acceptable model is at hand, we have chosen to employ a scheme which renders our results model-independent. We did this by "splitting" simultaneously both the first and second photons of an atomic cascade. We viewed the light emitted on opposite sides of an assembly of excited atoms and focused it separately into two beams. The wavelength λ_1 on one side was selected to correspond to that of the first transition of the cascade, and that on the other, λ_2, to the second. The two light beams impinged on beam splitters, thus creating a total of four beams. Four associated photomultipliers labeled γ_{1A}, γ_{1B}, γ_{2A}, and γ_{2B} detected them. We monitored the coincidence rates between the four combinations: γ_{1A}-γ_{1B}, γ_{2A}-γ_{2B}, γ_{1A}-γ_{2B}, and γ_{2A}-γ_{1B}. A diagram of the arrangement is shown in Fig. 2.

Define $I_1(t)$ and $I_2(t)$ as the instantaneous intensity at the γ_{1A}-γ_{1B} beam splitter with wavelength λ_1, and that at the γ_{2A}-γ_{2B} beam splitter with wavelength λ_2, respectively. It follows directly from the Cauchy-Schwarz inequality that the following inequality holds:

$$\left[\int_{-T/2}^{T/2}\int_{-T/2}^{T/2}\langle I_1(t+t'+\tau_1)I_1(t+t''+\tau_1)\rangle dt'\,dt''\right]\left[\int_{-T/2}^{T/2}\int_{-T/2}^{T/2}\langle I_2(t+t'+\tau_2)I_2(t+t''+\tau_2)\rangle dt'\,dt''\right] \geq \left[\int_{-T/2}^{T/2}\int_{-T/2}^{T/2}\langle I_1(t+t'+\tau_1)I_2(t+t''+\tau_2)\rangle dt'dt''\right]^2 . \quad (11)$$

Using (3), we can write this as

$$C_{1A-1B}(0)C_{2A-2B}(0) \geq C_{1A-2B}(\tau)C_{1B-2A}(\tau) . \quad (12)$$

Here we have ignored a possible polarization dependence of the detectors, and the finite photocathode areas, as well as the nonvanishing phototube dark rates. It can be shown that the inequality (12) may be summed over these contributions without change of form. Thus it is fully general and holds for these cases as well. The coincidence rates C_{1A-2B} and C_{2A-1B} here are the nonvanishing cascade rates. The product of these sets a lower bound to the product of the anomalous rates C_{1A-1B} and C_{2A-2B}. Thus, CFT predicts a large anomalous coincidence rate satisfying (12). The prediction of QFT significantly violates this inequality, requiring no coincidences except those due to two-atom excitations.

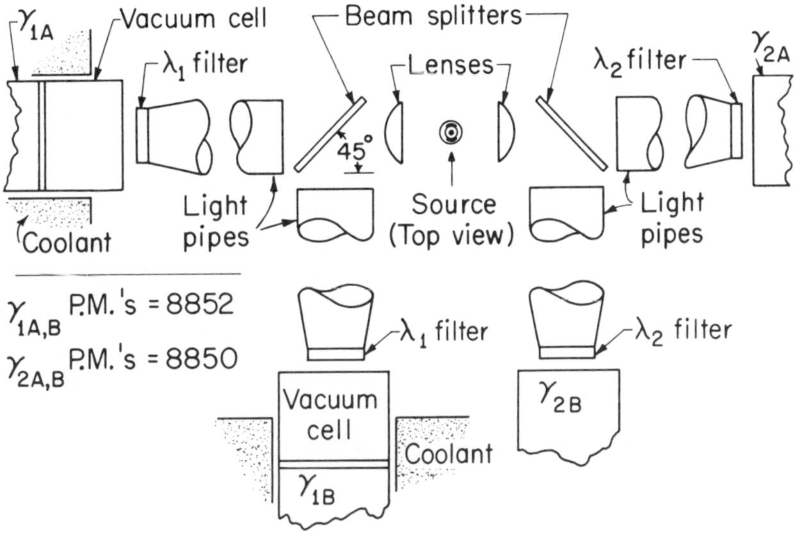

FIG. 2. Schematic diagram of our apparatus.

APPARATUS AND RESULTS

Figure 2 is a diagram of the apparatus. The source contained ^{202}Hg atoms which were excited by electron bombardment. Light produced at $\lambda_1 = 5676$ Å and $\lambda_2 = 4358$ Å by the cascade $9\,^1P_1 \rightarrow 7\,^3S_1 \rightarrow 6\,^3P_1$ was used. It was made parallel by lenses (aspheric, $f \approx 1$), and fell on TiO_2-coated glass beam splitters (transmission ≈ 63% and 35% for opposite linear polarizations, inclined at 45° to the incident beams). Each resulting beam was directed through an interference filter [transmission ≈ 50% at 5676 Å, full width at half maximum (FWHM) ≈ 50 Å for γ_{1A} and γ_{1B}; transmission ≈ 30% at 4358 Å, FWHM ≈ 100 Å for γ_{2A} and γ_{2B}] onto an appropriate photomultiplier tube [RCA 8852, quantum efficiency (QE) ≈ 15% at 5676 Å, dark current ≈ 50–300 Hz, operated at −80 °C for γ_{1A} and γ_{1B}; RCA 8850, QE ≈ 30% at 4358 Å, dark current ≈ 100 Hz, operated at 20 °C for γ_{2A} and γ_{2B}].

The source itself was patterned after a design by Holt, Nussbaum, and Pipkin,[18] and was made by using standard techniques. The electron gun was a standard 10-W cathode-ray-tube gun obtained through the courtesy of the Raytheon Corporation. It was mounted with suitable deflecting electrodes and light masks in a quartz and Pyrex envelope, evacuated, and cleaned by baking and discharging; the metal parts were outgassed by induction heating, and the oxide cathode was activated. A few milligrams of 93%-pure ^{202}Hg were then distilled into the tube and the envelope sealed. The Hg vapor pressure was controlled by keeping a side arm immersed in ice water. A beam current of approximately 0.7 μA traversed the cylindrical excitation region (length ≈ 2 mm, diam. ≈ 1 mm). The light output was stable. Photomultipliers operating in coincidence were separated from each other by more than 1.5 m to eliminate anomalous coincidences caused by cosmic rays. Light pipes minimized the light loss during transit. The interference filters were placed at the outer ends of the light pipes to minimize anomalous coincidences due to scintillations in the beam splitter and collimator lenses. These could be excited by cosmic rays and/or residual radioactivity therein. This configuaration also effectively eliminated phototube crosstalk induced by light emitted at the last dynodes. High-speed electronics with ≈ 1-nsec resolving time were used. The discriminators drove a time-to-amplitude converter whose output was fed to a pulse-height analyzer. External slow coincidence circuits gated the signals into one of the four analyzer memory quadrants, corresponding to the particular coincidence mode. The analyzer thus simultaneously accumulated the four different delayed coincidence spectra, i.e., the number of events pairs as a function of event separation time.

The results, shown in Figs. 3(a)–3(d), represent more than 26 hours of integration. We find no evidence for an anomalous coincidence rate in either the γ_{1A}-γ_{1B} or the γ_{2A}-γ_{2B} mode, but the normal cascade mode is quite apparent. For a timing and sensitivity check, both tube pairs were excited through the beam splitters by short-duration

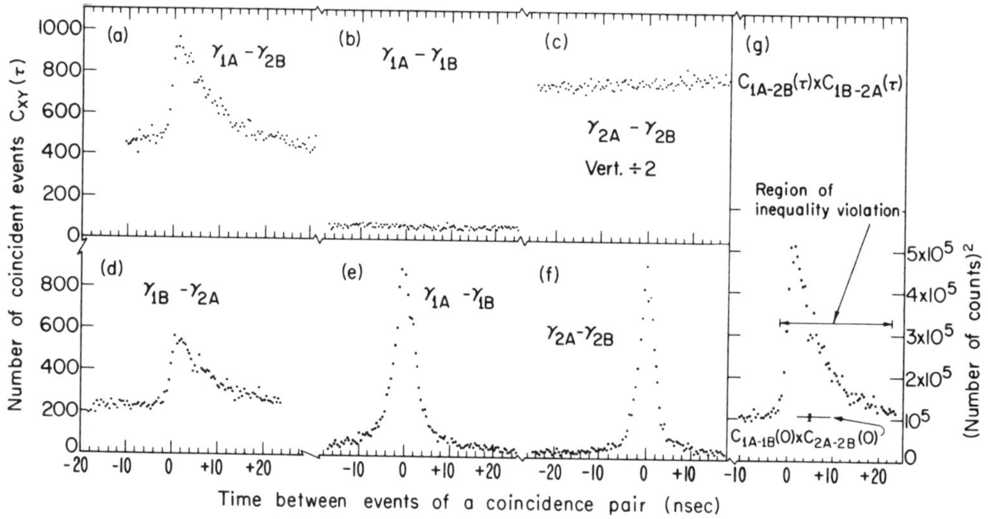

FIG. 3. (a)–(d) Time-delay coincidence spectra of the four monitored channels: C_{1A-2B}, C_{1A-1B}, C_{2A-2B}, and C_{1B-2A}. (e)–(f) C_{1A-1B} and C_{2A-2B} coincidence spectra in response to short pulses of light incident upon beam splitters produced by a barium titanate source. (g) Product of C_{1A-2B} and C_{1B-2A} versus time delay. For small times this clearly exceeds the indicated value of the product C_{2A-2B} and C_{1A-1B} evaluated at zero delay.

"classical" light pulses from a barium-titanate source,[19] with approximately one photon per pulse. The resultant coincidence spectra are shown in Figs. 3(e) and 3(f). Finally, Fig. 3(g) shows that our data severely violate (12) for a wide range of delays τ.

DISCUSSION

The importance of experimentally demonstrating phenomena which require a quantization of the electromagnetic field has been emphasized recently by a number of suggestions that such a quantization is unnecessary. Many standard effects have thus been challenged as not providing definitive proof for the necessity of this quantization.[4,20] Several recent experiments testing the specific predictions of NCT and the Schrödinger interpretation have been performed[21] in this direction. The present experiment and others[20] have tested the quantum-mechanical aspects of Maxwell's equations. So far, none has uncovered any departure from the quantum-electrodynamic predictions, but severe departures from CFT predictions have been found. The classical (unquantized) Maxwell equations thus appear to have only limited validity.

ACKNOWLEDGMENTS

The author thanks J. A. Crawford and M. H. Prior for helpful and stimulating discussions during the performance of this experiment.

---

*Work supported by the U. S. Atomic Energy Commission.

[1]See, for example, W. Davis and L. Mandel, in *Coherence and Quantum Optics: Proceedings of the 3rd Rochester Conference on Coherence and Quantum Optics*, edited by L. Mandel and E. Wolf (Plenum, New York, 1973), p. 113 and references therein.

[2]See, for example, W. Heitler, *Quantum Theory of Radiation*, 3rd edition (Oxford, London, 1954), p. 58.

[3]L. Mandel, E. C. G. Sudarshan, and E. Wolf, Proc. Phys. Soc. Lond. 84, 435 (1964); W. E. Lamb and M. O. Scully, in *Polarization: Matiere et Rayonnement*, edited by Société Française de Physique (Presses Universitaires de France, Paris, 1969).

[4]M. D. Crisp and E. T. Jaynes, Phys. Rev. 179, 1253 (1969); 185, 2046 (1969); C. R. Stroud, Jr. and E. T. Jaynes, Phys. Rev. A 1, 106 (1970); E. T. Jaynes; *ibid.* 2, 260 (1970). See also the extensive review by E. T. Jaynes, in *Coherence and Quantum Optics* (see Ref. 1), p. 35, and references therein.

[5]A. Ádám, L. Jánossy, and P. Varga, Acta Phys. Hung. 4, 301 (1955); Ann. Physik 16, 408 (1955).

[5a]Similar experiments were also performed by E. Brannen and H. I. S. Ferguson [Nature (Lond.) 178, 481 (1956)], and earlier with x rays by M. P. Givens [Philos. Mag. 37, 335 (1946)].

[6]J. M. Jauch, in *Foundations of Quantum Mechanics, International School of Physics "Enrico Fermi,"* edited by B. d'Espagnat (Academic, New York, 1971), p. 20; J. M. Jauch, *Are Quanta Real?* (Indiana Univ. Press, Bloomington, Indiana, 1973).

[7]L. Jánossy and Zs. Náray, Acta Phys. Hung. 7, 403 (1957).

[8]Y. Aharonov, D. Falkoff, E. Lerner, and H. Pendleton, Ann. Phys. (N.Y.) 39, 498 (1966).

[9]U. M. Titulaer and R. J. Glauber, Phys. Rev. 140, B676 (1965).

[10]E. Fermi, Rev. Mod. Phys. 4, 87 (1932).

[11]U. Fano, Am. J. Phys. 29, 539 (1961).

[12]An additional coincidence rate, the Brown-Twiss effect, is also due to two-atom excitations (see Ref. 11). Its contribution to large-aperture–large-solid-angle systems such as those under discussion here is negligible.

[13]A. T. Forester, R. A. Gudmundsen, and P. O. Johnson, Phys. Rev. 99, 1691 (1955).

[14]C. Camhy-Val and A. M. Dumont, Astron. and Astrophys. 6, 27 (1970).

[15]L. Mandel and E. Wolf, Rev. Mod. Phys. 37, 231 (1965). See also *Selected Papers on Coherence and Fluctuations of Light*, edited by L. Mandel and E.Wolf (Dover, New York, 1970), Vols. I and II.

[16]L. Mandel, Proc. Phys. Soc. Lond. 72, 1037 (1958). More recent discussions are summarized by J. R. Klauder and E. C. G. Sudarshan, *Fundamentals of Quantum Optics* (Benjamin, New York, 1968), Chap. 3.

[17]One might assume a model in which all decays are identical, with the emitted intensity nearly isotropic. On the other hand, recent direct observations of atomic recoil associated with spontaneous emission indicate that, in a semiclassical scheme, either the radiation must be emitted with a preferred direction, or this recoil is due to some other mechanism [J. L. Picque and J. L. Vialle, Opt. Commun. 5, 402 (1972); R. Schieder, H. Walther, and L. Wöste, *ibid.* 5, 337 (1972); R. Frisch, Z. Physik 86, 42 (1933)]. These experiments were in response to an elegant proof by A. Einstein [Phys. Z. 18, 121 (1917)] showing that recoil is necessary if thermal equilibrium is to be maintained when a dilute gas interacts with radiation. (His arguments also relied upon microscopic energy conservation.) Various schemes to account for such beaming have been proposed. See C. W. Oseen, Ann. Physik 69, 202 (1932) and the more recent work by Y. Beers, Am. J. Phys. 40, 1139 (1972); 41, 275 (1973).

[18]R. A. Holt, Ph.D. thesis, Harvard University, 1973 (unpublished); G. H. Nussbaum and F. M. Pipkin, Phys. Rev. Lett. 19, 1089 (1967).

[19]T. G. Innes and G. C. Cox, in Lawrence Berkeley Laboratory Counting Handbook, LBL Report No. UCRL-3307 Rev. (unpublished), pp. cc 8–31.

[20]J. F. Clauser, Phys. Rev. A 6, 49 (1972).

[21]J. M. Wessner, D. K. Anderson, and R. T. Robiscoe,

Phys. Rev. Lett. 29, 1126 (1972); H. M. Gibbs, *ibid.* 29, 459 (1972) and in *Coherence and Quantum Optics* (see Ref. 1), p. 83; H. M. Gibbs, G. G. Churchill, and G. J. Salamo, Phys. Rev. A 7, 1766 (1973); E. Gaviola, Nature 12, 772 (1928). This last experiment is discussed by J. M. Wessner, D. K. Anderson, and R. T. Robiscoe in Phys. Today 26 (No. 2), 13 (1973).